Green Synthesized Iron-based Nanomaterials

This book covers different green approaches used for the preparation of iron-based nanoparticles and their environmental remediation applications. It investigates various environmental applications such as antimicrobial studies, dye and heavy metal removal, and wastewater pollutant degradation by various green synthesized iron nanomaterials. Critical analysis of different routes and the separation techniques of iron-based nanomaterials along with the potential impacts of iron nanoparticles on human health and the atmosphere are also discussed. Overall, the authors:

- Summarize all the basic aspects of synthesis and application of iron-based green nanomaterials
- Explain morphological parameters of the prepared nanomaterials obtained from different routes and their specific applications
- Include different separation techniques from an industrial point of view
- Exclusively cover specific environmental remediation applications
- Discuss the future scope and challenges of green processes.

This book is aimed at researchers and professionals in chemical and environmental engineering, materials science, and nanotechnology.

Emerging Materials and Technologies

Series Editor
Boris I. Kharissov
Zongyu Huang, Xiang Qi, and Jianxin Zhong

Thin Film Coatings: Properties, Deposition, and Applications
Fredrick Madaraka Mwema, Tien-Chien Jen, and Lin Zhu

Biosensors: Fundamentals, Emerging Technologies, and Applications
Sibel A. Ozkan, Bengi Uslu, and Mustafa Kemal Sezgintürk

Error-tolerant Biochemical Sample Preparation with Microfluidic Lab-on-chip
Sudip Poddar and Bhargab B. Bhattacharya

Geopolymers as Sustainable Surface Concrete Repair Materials
Ghasan Fahim Huseien, Abdul Rahman Mohd Sam, and Mahmood Md. Tahir

Nanomaterials in Manufacturing Processes
Dhiraj Sud, Anil Kumar Singla, and Munish Kumar Gupta

Advanced Materials for Wastewater Treatment and Desalination: Advanced Materials for Wastewater Treatment and Desalination
A.F. Ismail, P.S. Goh, H. Hasbullah, and F. Aziz

Green Synthesized Iron-based Nanomaterials: Applications and Potential Risks
Piyal Mondal and Mihir Kumar Purkait

Polymer Nanocomposites in Supercapacitors
Soney C George, Sam John, and Sreelakshmi Rajeevan

For more information about this series, please visit: https://www.routledge.com/Emerging-Materials-and-Technologies/book-series/CRCEMT

Green Synthesized Iron-based Nanomaterials

Applications and Potential Risks

Piyal Mondal

and

Mihir Kumar Purkait

CRC Press is an imprint of the
Taylor & Francis Group, an **informa** business

First edition published 2023
by CRC Press
6000 Broken Sound Parkway NW, Suite 300, Boca Raton, FL 33487-2742

and by CRC Press
4 Park Square, Milton Park, Abingdon, Oxon, OX14 4RN

CRC Press is an imprint of Taylor & Francis Group, LLC

ISBN: 978-1-032-15326-1 (hbk)
ISBN: 978-1-032-15327-8 (pbk)
ISBN: 978-1-003-24363-2 (ebk)

DOI: 10.1201/9781003243632

Typeset in Times
by SPi Technologies India Pvt Ltd (Straive)

Contents

Preface

This book contains the various green synthesis routes of iron and iron-based nanomaterials, their environmental remediation applications, and their potential risks to the environment. Recently, green route synthesis of metal-based nanoparticles has attracted a lot of interest due to its low cost, nontoxic effects on the environment, and its unique physical and chemical properties. Iron, being abundant in nature, inexpensive, and having magnetic properties has been used in its nanoparticle form for numerous environmental applications. This work was utilized for preparing iron-based nanomaterial, and covers different green route approaches such as: plant source extract, microorganism based, biocompatible green reagents (such as cellulose, tannic acid, ascorbic acid, hemoglobin, etc.), microwave synthesis, hydrothermal synthesis, and plant wastes. The book also provides a comprehensive investigation of various environmental applications such as: antimicrobial study, dye removal, heavy metal removal, and wastewater pollutant degradation by various green synthesized iron nanomaterials. Recent developments regarding green synthesis of iron-based bimetallic nanocomposites and their advantages over iron NPs are a prime focus. Critical analysis of different routes and the separation study of iron-based nanomaterial from an industrial point of view are an added advantage of this book. Further, this book addresses various research possibilities and queries of postgraduate level students, academics, and researchers of various fields.

The motivation behind writing the book was to provide a single, focused, and dedicated source on the different green synthesis routes followed to date in preparing iron-based nanomaterials, their potential environmental applications, and their future scope and challenges. It focuses on the recent trends and techniques utilized for preparing nontoxic, environmental-friendly, iron-based nanoparticles with a detailed, efficient investigation of their morphological structures, shape, and particular environmental applications. The book covers critical analysis of the mentioned green routes in terms of their stability, scale-up properties, production costs, and agglomeration prevention. Moreover, the basic problem of separating the utilized iron-based nanomaterial after applications is also discussed. The book serves as a unique and complete guide for researchers on preparing green synthesized iron-based nanomaterials, their specific morphology, along with their probable environmental applications. It widens the scope for future research directions with specific challenges to overcome.

Authors

Dr. Piyal Mondal received his BTech in Chemical Engineering from the National Institute of Technology Durgapur, West Bengal, India during 2012. He completed his master's degree and PhD in chemical engineering at the Indian Institute of Technology, Guwahati, India. His research work is dedicated to preparing various surface engineered polymers for specific environmental applications. The synthesis of polymeric membranes, green synthesized nanomaterials, and hybrid techniques to combat wastewater treatment are also his research focus. He has fabricated different prototypes for environmental separation applications. Currently, he has authored five reference books namely: Stimuli-responsive Polymeric Membrane (Elsevier, ISBN: 9780128139615), Treatment of Industrial Effluents (CRC Press, ISBN: 9780429401763), Thermal Induced Membrane Separation Processes (Elsevier, ISBN: 9780128188019), Hazards and Safety in Process Industries: Case Studies (CRC Press, ISBN 9780367516512), pH Responsive Membranes: Biomedical Applications (CRC Press, ISBN 9781032061672), and a few more are in progress. Moreover, his publications consist of 17 peer-reviewed articles in reputed international journals, with several more under review. He has presented more than ten papers and received several awards in poster and paper presentations in his field at international and national conferences.

Dr. Mihir Kumar Purkait is a professor in the Department of Chemical Engineering and Head Centre for the Environment at the Indian Institute of Technology, Guwahati (IITG). Prior to joining the faculty at IITG in 2004, he received his PhD and M. Tech in chemical engineering from the Indian Institute of Technology, Kharagpur (IITKGP), after completing his BTech and BSc (Hons) in chemistry at the University of Calcutta. He has received several awards including; Dr. A.V. Rama Rao Foundation's Best PhD Thesis and Research Award in Chemical Engineering from IIChE (2007); the BOYSCAST Fellow Award (2009–2010) from the DST; the Young Engineer's Award in the field of chemical engineering from the Institute of Engineers (India, 2009); and the Young Scientist Medal Award from the Indian National Science Academy (INSA, 2009). Professor Purkait is a Fellow of the Royal Society of Chemistry (FRSC) UK, and a Fellow of the Institute of Engineers (FIE) India. He is the director of two incubated companies (RD Grow Green India Pvt. Ltd. and Vixudha Bio Products Ltd.). He is also technical advisor of Gammon India Ltd and the Indian Oil Corporation, Bethkuchi for their treatment plant. His current research activities are focused in four distinct areas: i) advanced separation technologies; ii) waste to energy; iii) smart materials for various applications; and iv) process intensification. In each of these areas, his goal is to synthesize stimuli responsive materials and to develop a more fundamental understanding of the factors governing the performance of the chemical and biochemical processes. He has more than 20 years of academic experience and research and has published more than 200 papers in different reputed journals (Citation: >9500, h-index = 64, 10 index =151). He has eight patents and has completed twenty-four sponsored and consultancy projects from various funding agencies. Professor Purkait has guided eighteen PhD students and authored six books.

1 Green Synthesis of Iron Nanomaterials and its Mechanism

1.1 INTRODUCTION

Nanotechnology is defined as the transformation of matter using physical, chemical, and biological techniques and processes, with or without incorporating other ingredients, to produce substances with specialized functionalities, enhanced characteristics and specific attributes, which can be used in diverse field of applications (Goswami et al., 2012; Taghizadeh et al., 2013). A nanoparticle (NP) can be defined as a minuscule particle which has at least one of its size/dimension, not more than 100 ηm (Thakkar et al., 2010). The nanoparticles thus produced are distinct from the bulk materials, as they exhibit unique electrochemical, optical, and thermal properties, along with a substantially larger surface area to volume ratio (Taghizadeh et al., 2013). These unique properties of the nanoparticles are responsible for their popularity and widespread use in the fields of agriculture, biotechnology, chemistry, communications, consumer goods, defense, electronics, energy, environmental remediation, heavy industries, material science, medicine, microbiology, optics, and various engineering fields (Asfaram et al., 2017; Mokhtari et al., 2016).

Metals and their corresponding oxides are extensively converted into nanoparticles using both physical and chemical processes (Ghaedi et al., 2013; Raul et al., 2012). Chemical synthesis methods, like element lowering and the sol serum approach, intensively use toxic and hazardous substances such as sodium borohydride, hypophosphite, and hydrazine hydrate, which harm the environment (Changmai et al., 2017; Goswami and Purkait, 2014). So, the development of a competent, cost-efficient, and sustainable green process for the synthesis of nanoparticles is an ongoing effort. Stable and well-functionalized nanoparticles have been produced from organisms like bacteria, actinomycetes, fungi, yeast, viruses, and other microbes (Mandal et al., 2006). These microorganisms serve as environment-friendly and sustainable precursors. However, factors like local availability of resources, social adaptability, and economic feasibility heavily influence the overall sustainability of the process (Dey and Purkait, 2015). Apart from metal NPs, metal-based nanocomposites have also been extensively utilized for various environmental hazard purposes (Ghaedi et al., 2016).

DOI: 10.1201/9781003243632-1

Nanoparticle synthesis employing green chemistry techniques has several advantages over the conventional methods: it is safer, does not contaminate the environment, and is relatively inexpensive. Gonzalez-Moragas et al. (2015) prepared superparamagnetic iron oxide nanoparticles (SPIONs) of size 5.9 ± 1.4 nm through a microwave-assisted, commercial chemical process using iron (III) acetylacetonate as a precursor, with a production cost of €130 per 10 g of SPION. Whereas the cost for preparing green synthesized iron NPs from plant extract is around USD 0.5–10 per gram, depending upon species and purity (Bolade et al., 2019). The advantage of such green synthesized NPs is that it does not require synthetic reducing agents which are environmentally hazardous. Moreover, the process does not require external capping and stabilizing agents. Commercially available Fe NPs have a diverse price spectrum depending on factors like oxidation state, particle size, functionalization, and a dispersion medium which can range from USD 50–500 per gram (Sigma-Aldrich, St. Louis, MO, USA). Various conventional chemical techniques show the disadvantages of high cost (Goswami and Purkait, 2015). Other advantages and salient features of green synthesis processes are first, the dual use of the active natural component, like the extract, as both reducing and capping agent. This ability enables the production of small size nanoparticles in large scale units (Ghaedi et al., 2015). Second, the time taken for reducing the metal ions is short, and the product subsequently formed is immensely stable due to the presence of stabilizing agents in the extract (Mondal and Purkait, 2019). Last but not least, a plethora of both microscopic and macroscopic organisms are available that could be utilized for the desired nanoparticle functionality (Mohanpuria et al., 2008).

Iron nanoparticles (Fe NPs) exhibit superb dimensional stability, nontoxicity and have an affinity to form oxides. Apart from having a high surface area, electrical and thermal conductivity, Fe NPs also possess magnetic properties and are known as magnetic nanoparticles (Arabi et al., 2016). Superparamagnetism is a unique consequence of magnetic nanoparticles. It is observed in ferromagnetic or ferrimagnetic nanomaterials, only when the size and number of domains, are both adequately small, broadly between 10 and 150 nm in diameter depending on the material (Clemons et al., 2019). At these extreme sizes, they become a single domain magnetic material that has no hysteresis loop; also, their magnetization can randomly flip direction under the influence of temperature (Frenkel and Doefman, 1930). Iron nanoparticles that exhibit superparamagnetism are termed as superparamagnetic iron oxide nanoparticles (SPIONs), and find wide application in the biomedical field.

Iron is abundant in nature, less expensive, and its magnetic behavior has engrossed synthesis of NPs having enormous scope in environmental separation applications. However, it is envisaged that NPs of various metal and metal oxides such as silver, gold, copper, and zinc should be studied as thoroughly that of iron NP. A few review papers have concentrated on specific green methods using sources from plant parts (Bolade et al., 2019; Fahmy et al., 2018). In addition to this, the present review article discusses the state of artwork on the green synthesis of iron NPs and iron nanocomposites using various biocompatible green reagents such as biopolymers, cellulose, hemoglobin, and glucose, along with microorganisms and low cost energy related processes. Again, iron-based nanocomposites through green routes and their environmental application are scarce in the recent literature and are also elaborated here. The tables provide a detailed insight into the morphological structure of iron nanomaterials/nanocomposites prepared from each specific source, along with their specific applications and efficiency. The readers of this book will benefit from up-to-date information and critical expert suggestions on the green synthesis of iron nanoparticles/nanocomposites and their enormous potential in environmental separation and biomedical applications, along with some potential difficulties.

1.2 GREEN ROUTE SYNTHESIS OF IRON NANOPARTICLES (Fe NPs)

An efficient and eco-friendly methodology to synthesize nanoparticles should ideally use renewable energy, minimize waste discharge, and optimize energy consumption according to the principles of green chemistry (Anastas and Werner, 1998). Thus, it may be conceptualized that the green synthesis process must utilize plants, microbes, biopolymers, and waste materials for the active biocomponent along with low heating requirements and benign solvents (Bolade et al., 2019). Water,considered as the universal solvent, has been reported in many studies as a solvent with biomolecules, mostly polyphenols, extracted from plants as reducing, capping, and stabilizing agents. These bioactive polyphenols have been extracted from leaves, stems, roots, flowers, fruits, fruit peels, seeds, gums, and even wastes of various vegetations (Yew et al., 2020). Agglomeration is the main problem in chemical synthesis of NPs, but green route techniques have quite significantly limited that aspect. Especially for plant extract and microorganism mediated synthesis, the polyphenolic content and natural proteins act as stabilizer and coating agent over the prepared NPs, which restricts the agglomeration. Most of the prepared NPs through green route techniques are generally stored in aqueous form at definite pH to provide stability and uniform particle size (Bolade et al., 2019). Better stabilization can also be obtained by utilizing biocompatible reagents and other chemical surface coatings, which provide less reactivity when coming into contact with environmental fluids. NPs stored in powder form should be kept in a vacuum otherwise agglomeration occurs in the presence of moisture from the environment.

Prabhakar and Samadder (2017) have synthesized Fe NPs using extracts from both aquatic and terrestrial weeds and explored their application in eutrophic wastewater treatment for the removal of nitrates and phosphates. Production of toxic by-products and their subsequent disposal with other harmful reagents can be reduced substantially by using aqueous media and ambient temperature and pressure. Figure 1.1 represents the nano-structured Fe NPs prepared from the green source.

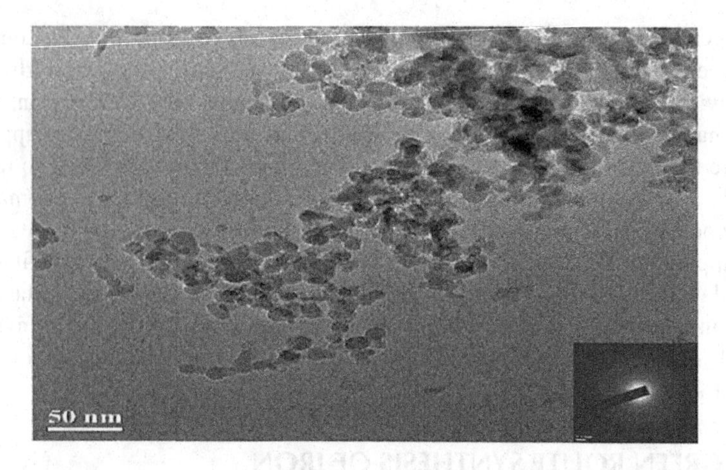

FIGURE 1.1 Transmission Electron Microscopy (TEM) micrographs of synthesized iron nanoparticles using *Eichhornia crassipes* [Reproduced with permission from Prabhakar and Samadder (2017) © Elsevier].

Time for synthesis and energy consumption can be reduced by incorporating a microwave heating source instead of conventional methods. The microwave-assisted technique undergoes hydrothermal process and has thus been widely used by various researchers for scaling up iron NPs synthesis purpose.

Gonzalez-Moragas et al. (2015) adopted the microwave-based thermal technique for synthesizing SPIONs at a rate of 3g h^{-1} with batch mode utilizing multimode microwave unit. The study implemented a multivessel rotor system, which may help to overcome the production bottleneck associated with single-mode microwave systems. Figure 1.2 represents the microscopic structures of the scale-up synthesized Fe NPs.

Moreover, Bowman et al. (2008) utilized both batch and continuous flow models to synthesize NPs which provided yield of several multigrams per hour. Production of NPs in the range of kg or ton per day has not yet been reported. However, scaled versions of production of such NPs can be enhanced with a continuous flow type reactor aided with a multimode microwave system. Iron nanoparticle formation, facilitated by microorganisms like actinomycetes, algae, fungi, and yeast at neutral pH and close to room temperature, is extensively reported (Schröfel et al., 2014; Thakur et al., 2014). Catalytic, biological, and environmental applications of iron nanoparticles synthesized via green routes are widely reported (Ahmed et al., 2014).

Chapter 2 lists all the biocompatible reagents utilized for synthesizing iron NPs along with iron-based nanomaterials, their composites, and bimetallic nanocomposites. Synthesis involving microorganisms, various plant parts, plant waste materials, microwave-assisted synthesis, and hydrothermal processes are discussed in depth in Chapters 3 through 6 to provide an overview of green synthesized Fe and Fe-based nanomaterials. Moreover, this book focuses on the

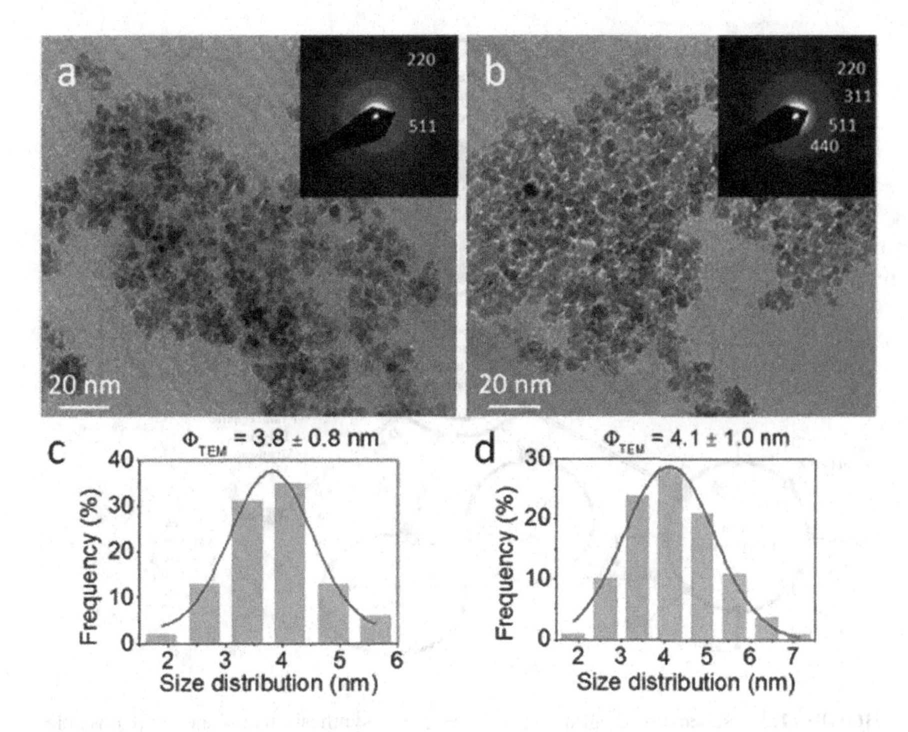

FIGURE 1.2 Characterization of the scale-up batches by TEM. Panels a and b show TEM images of: (a) SU-1 (3.8 ± 0.8 nm, at normal scale-up condition), (b) SU-2 (4.1 ± 1.0 nm, extended after scale-up condition). Panels c and d show the size distribution histogram of the scale-up batches: (c) SU-1, (d) SU-2 [Reproduced with permission from Gonzalez-Moragas et al. (2015) © Elsevier].

various iron-related bimetallic nanocomposites, which have been found to have better catalytic and adsorptive properties in comparison to iron-only NPs. Various environmental applications are discussed in Chapter 7 related to dye degradation, heavy metal removal, fluoride removal along with wastewater treatment, and other waterborne pollutant removals. Chapter 8 focuses on the various biomedical applications of iron NPs apart from environmental remediation applications.

Figure 1.3 gives an overall schematic presentation of the discussion in this book related to iron and iron-based nanoparticles. The figure explains all the possible green synthesis routes utilized to date for preparing iron nanoparticles along with the new inclusion of their derivatives with specific applications. Moreover, the hazards and toxicity of such green synthesized nanomaterials on living organisms have been elaborated in Chapter 10, with regard to direct contact with soil and water. Future challenges have been outlined with a suitable scope of improvement in preparation and proper management of iron-based NPs in later chapters of this book.

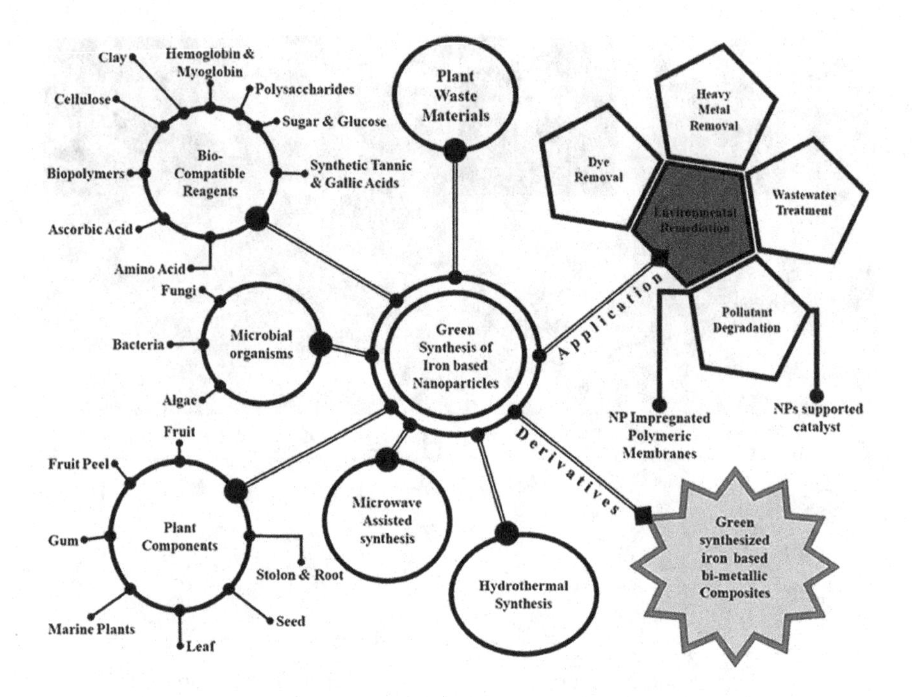

FIGURE 1.3 Schematic diagram of various green synthesis techniques and possible environmental applications of iron, and iron-based nanomaterial [Reproduced with permission from Mondal et al. (2020) © Elsevier].

1.3 MECHANISM OF NANOPARTICLE SYNTHESIS

Broadly, NPs are synthesized by two dominant approaches: the first is the bottom-up approach (BUA), and the second is the top-down approach (TDA) (Yang et al., 2019). In the BUA, NP synthesis is carried out using biopolymers and smaller active compounds present in plant extract and microorganism secretions (Calzoni et al., 2019). Conversely, in the TDA, the bulk of NPs are size-reducing products of various photochemical reactions during the bottom-down approach, which further breaks into nanosized particles. For understanding the synthesis approaches, a graphical view is represented in Figure 1.4. Both reaction mechanisms follow the same chemistry of reduction and oxidation as shown in Figure 1.5. In general, the perfection of homogeneity and fewer defects are observed in many biogenic reactions, i.e., BUA (Yan et al., 2019). Further, a plant extract's active compounds serve as a capping agent that stabilizes NPs for a long time (Veisi et al., 2015).

The plant-based green synthesis technique is widely adopted compared to the others. The plants are easy to find and biologically safer than the microorganism and fungi-mediated synthesis. Both involve a lengthy process of cell culture (Khan et al., 2019; Ahmed and Khalil, 2020). Plant extract contains active compounds of terpenoids, alkaloids, phenols, tannins, and vitamins known for their therapeutic and environmental values (Savoia, 2012). These compounds' basic

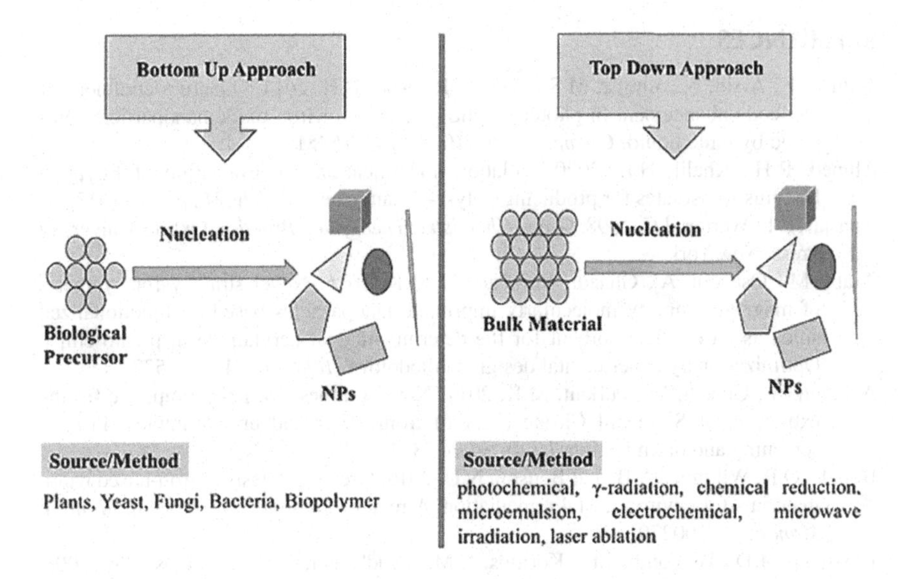

FIGURE 1.4 Graphical representation of techniques used in NP synthesis [Reproduced with permission from Qamar and Ahmad (2021) © Elsevier].

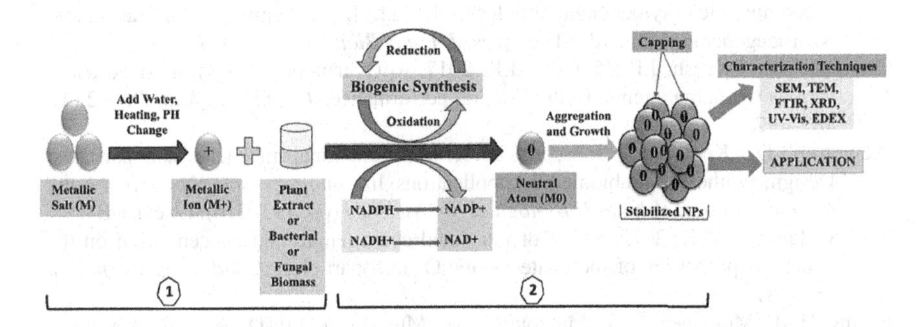

FIGURE 1.5 Summary for the biosynthesis of nanoparticles (1) Biomolecules are found in plant extract or secreted by bacteria or fungi and act as capping and reducing agents. These molecules are sugar, carbohydrate, enzymes, and proteins, which reduce metallic ions from (M^+) to (M^0) via oxidation/reduction mechanism. (2) Reduced metallic form aggregates and creates clusters of NPs that can be confirmed through the reaction mixture by changing color [Reproduced with permission from Qamar and Ahmad (2021) © Elsevier].

building blocks are simple hydrocarbon, monoterpenoids, isoprenoids, carotenoids, sesquiterpenoids, turpentine, and sterol. Metabolites of these small molecular blocks form complex metabolites such as phytol tail in chlorophyll (Mashwani et al., 2016). Due to the presence of these valuable compounds, NPs synthesized from plants are more stable. These compounds can cap NPs, crucial for aggregation, inhibition, and growth termination. In addition, they are determinants of the size and shape of NPs due to the difference in reducing ability.

REFERENCES

Ahmed, F., Arshi, N., Anwar, M.S., Danish, R., Koo, B.H. 2014. Quantum-confinement induced enhancement in photocatalytic properties of iron oxide nanoparticles prepared by ionic liquid. *Ceram. Int.* 40 (10), 15743–15751.

Ahmed, R.H., Khalil, H.B. 2020. Isolation, identification and evaluation of Egyptian Bacillus sp. isolates for producing poly-γ-glutamic acid. *J. Adv. Micro.* 10, 1–13.

Anastas, P.T., Werner, J.C. 1998. *Green Chemistry: Theory and Practice*. Oxford University Press, New York.

Arabi, M., Ostovan, A., Ghaedi, M., Purkait, M.K. 2016. Novel strategy for synthesis of magnetic dummy molecularly imprinted nanoparticles based on functionalized silica as an efficient sorbent for the determination of acrylamide in potato chips: Optimization by experimental design methodology. *Talanta* 154, 526–532.

Asfaram, A., Ghaedi, M., Purkait, M.K. 2017. Novel synthesis of nanocomposite for the extraction of Sildenafil Citrate (Viagra) from water and urine samples: Process screening and optimization. *Ultra. Sono.* 38, 463–472.

Bolade, O.P., Williams, A.B. and Benson, N.U. 2019. Green synthesis of iron-based nanomaterials for environmental remediation: A review. *Environ. Nanotechnol. Monitor. Manag.* 13, 100279.

Bowman, M.D., Holcomb, J.L., Kormos, C.M., Leadbeater, N.E., Williams, V.A. 2008. Approaches for Scale up of Microwave Promoted Reactions. *Org. Proc. Res. Develop.* 12, 41–57.

Calzoni, E., Cesaretti, A., Polchi, A., Di Michele, A., Tancini, B., Emiliani, C. 2019. Biocompatible Polymer Nanoparticles for Drug Delivery Applications in Cancer and Neurodegenerative Disorder Therapies. *J. Func. Biomater.* 10 (1), 4.

Changmai, M., Priyesh, J.P., Purkait, M.K. 2017. Al_2O_3 nanoparticles synthesized using various oxidizing agents: Defluoridation performance. *J. Sci.: Adv. Mat. Dev.* 2 (4), 483–492.

Clemons, T.D., Kerr, R.H., Joos, A. 2019. Multifunctional magnetic nanoparticles: Design, synthesis, and biomedical applications. In *Comprehensive Nanoscience and Nanotechnology: Volume 3: Biological Nanoscience* (pp. 193–210). Elsevier BV.

Dey, A., Purkait, M.K. 2015. Effect of fatty acid chain length and concentration on the structural properties of the coated $CoFe_2O_4$ nanoparticles. *J. Ind. Eng. Chem.* 24, 181–187.

Fahmy, H.M., Mohamed, F.M., Marzouq, M.H., Mustafa, A.B.E.-D., Alsoudi, A.M., Ali, O.A., Mohamed, M.A., Mahmoud, F.A. 2018. Review of green methods of iron nanoparticles synthesis and applications. *Bio Nano Sci.* 8, 491–503.

Frenkel, J., Doefman, J. 1930. Spontaneous and induced magnetisation in ferromagnetic bodies. *Nature* 126 (3173), 274–275.

Ghaedi, M., Abbasi Larki, H., Marahel, F., Sahraei, R., Purkait, M.K. 2013. Synthesis and Characterization of Zinc sulfide nanoparticles loaded on activated carbon for the removal of Methylene Blue. *Environ. Prog. & Sustain. Ener. (AIChE)*. 32 (3), 535–542.

Ghaedi, M., Hassani, R., Dashtian, K., Shafie, G., Purkait, M.K., Dehghan, H. 2016. Adsorption of methyl red on to palladium nanoparticles loaded on activated carbon: Experimental design optimization. *Desal. Water Treat.* 57 (47), 22646–22654.

Ghaedi, M., Yousefinejad, M., Safarpoor, M., Zare Khafri, H. Purkait, M.K. 2015. Rosmarinus officinalis leaf extract mediated green synthesis of silver nanoparticles, and antimicrobial properties. *J. Ind. Eng. Chem.* 31, 167–172.

Gonzalez-Moragas, L., Yu, S.-M., Murillo-Cremaes, N., Laromaine, A., Roig, A. 2015. Scale-up synthesis of iron oxide nanoparticles by microwave assisted thermal decomposition. *Chem. Eng. J.* 281, 87–95.

Goswami, A., Purkait, M.K. 2014. Removal of fluoride from drinking water using Nanomagnetite aggregated schwertmannite. *J. Water Proc. Eng.* 1, 91–100.

Goswami, A., Purkait, M.K. 2015. Preparation and characterization of nanoporous schwertmannite for defluoridation of water. *Res. J Sci. Technol.* 2, 65–83.

Goswami, P., Raul, K., Purkait, M.K. 2012. Arsenic adsorption using copper(II) oxide nanoparticles. *Chem. Eng. Res. Des.* 90, 1387–1396.

Khan, T., Ullah, N., Khan, M.A., Mashwani, Z.-U.-R., Nadhman, A. 2019. Plant-based gold nanoparticles; a comprehensive review of the decade-long research on synthesis, mechanistic aspects and diverse applications. *Adv. Colloid Interface Sci.* 272, 102017.

Mandal, D., Bolander, M.E., Mukhopadhyay, D., Sarkar, G., Mukherjee, P. 2006. The use of microorganisms for the formation of metal nanoparticles and their application. *App. Microbio. Biotechnol.* 69 (5), 485–492.

Mashwani, Z.-U.-R., Khan, M.A., Khan, T., Nadhman, A. 2016. Applications of plant terpenoids in the synthesis of colloidal silver nanoparticles. *Adv. Colloid Interface Sci.* 234, 132–141.

Mohanpuria, P., Rana, N.K., Yadav, S.K. 2008. Biosynthesis of nanoparticles: Technological concepts and future applications. *J. Nano. Res.* 10 (3), 507–517.

Mokhtari, P., Ghaedi, M., Dashtian, K., Rahimi, M.R., Purkait, M.K. 2016. Removal of methyl orange by copper sulfide nanoparticles loaded activated carbon: Kinetic and isotherm investigation. *J. Mol. Liq.* 219, 299–305.

Mondal, P., Purkait, M.K. 2019. Preparation and characterization of novel green synthesized iron–aluminum nanocomposite and studying its efficiency in fluoride removal. *Chemosphere* 235, 391–402.

Mondal, P., Anweshan, A., Purkait, M.K. 2020. Green synthesis and environmental application of Iron-based nano-materials and nanocomposite: A review. *Chemosphere* 259, 127509.

Prabhakar, R., Samadder, S.R. 2017. Jyotsana Aquatic and terrestrial weed mediated synthesis of iron nanoparticles for possible application in wastewater remediation. *J. Clean. Prod.* 168, 1201–1210.

Qamar, S.R., Ahmad, J.N. 2021. Nanoparticles: Mechanism of biosynthesis using plant extracts, bacteria, fungi, and their applications. *J. Mol. Liquids* 334, 116040.

Raul, P.K., Devi, R.R., Umlong, I.M., Banerjee, S., Singh, L., Purkait, M. K. 2012. Removal of fluoride from water using iron oxide-hydroxide nanoparticles. *J. Nano Sci. Nano Technol.* 12, 3922–3930.

Savoia, D. 2012. Plant-derived antimicrobial compounds: Alternatives to antibiotics. *Future Microbiol.* 7, 979–990.

Schröfel, A., Kratošová, G., Šafařík, I., Šafaříková, M., Raška, I., Shor, L.M. 2014. Applications of biosynthesized metallic nanoparticles—A review. *Acta Biomaterialia* 10 (10), 4023–4042.

Taghizadeh, F., Ghaedi, M., Kamali, K., Sharifpour, E., Sahraie, R., Purkait, M.K. 2013. Comparison of nickel and/or zinc selenide nanoparticle loaded on activated carbon as efficient adsorbents for kinetic and equilibrium study of removal of Arsenazo (III) dye. *Pow. Technol.* 245, 217–226.

Thakkar, K.N., Mhatre, S.S., Parikh, R.Y. 2010. Biological synthesis of metallic nanoparticles. *Nanomed.: Nanotechnol. Biol. Med.* 6 (2), 257–262.

Thakur, V.K., Thakur, M.K., Raghavan, P., Kessler, M.R. 2014. Progress in green polymer composites from lignin for multifunctional applications: A review. *ACS Sustain. Chem. Eng.* 2 (5), 1072–1092.

Veisi, H., Ghorbani-Vaghei, R., Hemmati, S., Aliani, M.H., Ozturk, T. 2015. Green and effective route for the synthesis of monodispersed palladium nanoparticles using herbal tea extract (*Stachys lavandulifolia*) as reductant, stabilizer and capping agent, and their application as homogeneous and reusable catalyst in Suzuki coupling reactions in water. *Appl. Organomet. Chem.* 29 (1), 26–32.

Yan, C., Yu, T., Ji, C., Kang, D.J., Wang, N., Sun, R., Wong, C.-P., 2019. Tailoring highly thermal conductive properties of Te/MoS_2/Ag heterostructure nanocomposites using a bottom-up approach. *Adv. Electron. Mat.* 5, 1800548.

Yang, M.-S., Song, C., Choi, J., Jo, J.-S., Choi, J.-H., Moon, B.K., Noh, H., Jang, J.-W. 2019. Fabrication of diffraction gratings by top-down and bottom-up approaches based on scanning probe lithography. *Nanoscale* 11, 2326.

Yew, Y.P., Shameli, K., Miyake, M., Khairudin, N.B.B.A., Mohamad, S.E.B., Naiki, T., Lee, K.X. 2020. Green biosynthesis of superparamagnetic magnetite Fe_3O_4 nanoparticles and biomedical applications in targeted anticancer drug delivery system: A review. *Arab. J. Chem.* 13 (1), 2287–2308.

2 Biocompatible Reagents-based Green Synthesis

2.1 BIOPOLYMERS

The use of innocuous synthetic biocompatible substances for the preparation and stabilization of magnetic nanoparticle polymer composites is well documented in the literature. He and Zhao (2005) investigated starch stabilized bimetallic Fe/Pd nanoparticles in an aqueous medium. Starch is a hydrophilic polymer with around 20% amylose content, and is reported to significantly improve the stability and dispersion of the iron NPs. Moreover, Gao et al. (2008) prepared magnetic Fe_3O_4 nanoparticle using sodium alginate biopolymer via redox-based hydrothermal method from ferric chloride hexahydrate and urea as precursors. The resultant nanoparticles were of uniform spherical morphology with a mean diameter of 27.2 nm.

Figure 2.1 represents the TEM image and size distribution chart of the biopolymer coated iron NPs synthesized by Gao et al. (2008). Further, Jegan et al. (2011) pioneered the synthesis of a magnetite (Fe_3O_4) agar nanocomposite via coprecipitation of ferric and ferrous ions. The nanomaterial was well dispersed in the solution with size 50–200 nm. Figure 2.2 represents the typical SEM images of the synthesized Fe_3O_4 nanoparticles.

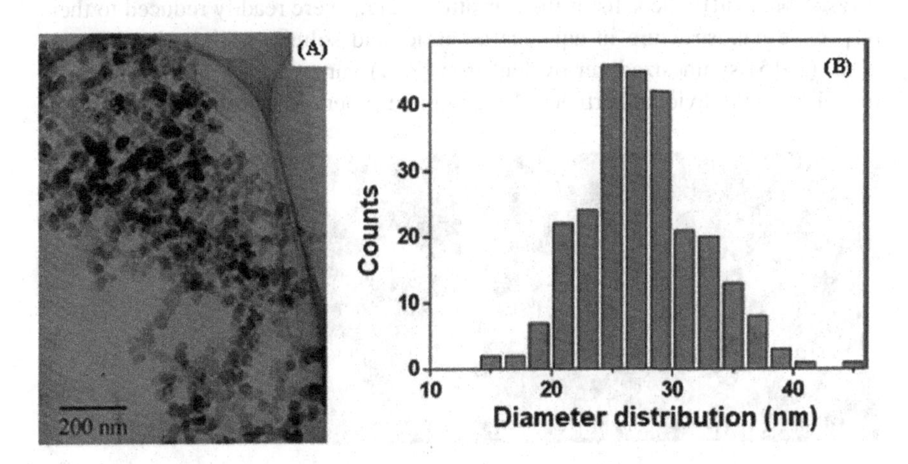

FIGURE 2.1 Typical TEM image (a) and size distribution histogram (b) of the synthesized iron NPs [Reproduced with permission from Gao et al. (2008) © Elsevier].

DOI: 10.1201/9781003243632-2

FIGURE 2.2 SEM images of prepared agar-Fe_3O_4 nanocomposite at different magnifications [Reproduced with permission from Jegan et al. (2011) © International Journal of Nano Dimension].

2.2 ASCORBIC ACID

Nadagouda and Varma (2007) studied the synthesis of Fe and Cu NP using Vitamin C (ascorbic acid). The salts of the transition metals were readily reduced to their respective nanostructure in aqueous ascorbic acid solution. Similarly, Savasari et al. (2015) synthesized zero-valent iron (ZVI) nanoparticles in ascorbic acid medium. The individual particles also exhibited spherical morphology, having an

FIGURE 2.3 The SEM images of (a) ascorbic acid stabilized ZVI nanoparticle and (b) non-stabilized ZVI nanoparticle [Reproduced with permission from Savasari et al. (2015) © Elsevier].

average particle size of 20–75 nm assembled in chain formation. Figure 2.3 shows the uniform distribution and agglomerated distribution of ZVI nanoparticles in presence of ascorbic acid and its absence respectively.

Furthermore, ascorbic acid has a stabilizing property and has been reported as a functionalizing agent. Ascorbic acid was utilized by Sreeja et al. (2014) to prepare, coat, and functionalize superparamagnetic iron oxide nanoparticles (SPION) with 5 nm diameters in a stable dispersion for medical application. The size and shape of the SPION were determined using a transmission electron microscope (TEM).

2.3 AMINO ACID

Krishna et al. (2012) employed the wet chemical coprecipitation method to synthesize amine-functionalized γ-Fe_2O_3 nanoparticles using L-Lysine (lys). The magnetization of γ-Fe_2O_3/lys magnetic NPs was 40.6 emu/g, with an average particle size of 14.5 nm. Likewise, a study by Siskova et al. (2013) reported the production of ZVI using various amino acids, namely: L-glutamic acid, L-glutamine, L arginine, and L-cysteine along with the effect of pH on ZVI yields.

2.4 HEMOGLOBIN AND MYOGLOBIN

Hemoglobin and myoglobin are naturally occurring biomolecules that contain iron. An investigation by Sayyad et al. (2012) reported the synthesis of Fe NPs stabilized at room temperature in a single-phase reduction reaction. The nanoparticles, thus produced, were crystalline with a narrow size band of 2–5 nm. This protocol is vital for the production of biocompatible NP required for medical applications. Figure 2.4 shows the SEM structure of the stabilized Fe NPs.

2.5 SUGAR AND GLUCOSE

Utilizing D-glucose as reducing agent, polycrystalline Fe_3O_4 nanoparticles were prepared by Lu et al. (2010) with gluconic acid as a stabilizer and dispersing agent. The prepared NPs were spherical in nature with average diameter of 12.5 nm. Furthermore, glucose and gluconic acid were utilized as capping agents by Sun et al. (2009) for fabricating magnetite nanoparticles by hydrothermal reduction process.

Moreover, Yan et al. (2015) utilized wood derived sugars and synthesized carbon encapsulated iron nanoparticles through hydrothermal carbonization process. Nanospheres with diameters of 100–150 nm were formed with an iron core of size 10–25 nm capable of catalytically converting syngas into liquid hydrocarbons. Figure 2.5 shows the SEM images of the encapsulated Fe NPs.

Demir et al. (2014) studied the effect of lactose, mannose, galactose, maltose, and fructose (different saccharides) on the synthesis of magnetite nanoparticles. All saccharides, except fructose, exhibited dual-purpose applicability as both a reducing agent and a capping agent. Due to its nonreducing nature, fructose acts only as a capping agent. The particle size distribution of synthesized Fe_3O_4 NPs was in the range of 3.8–13.1 nm, and the morphology of the nanoparticles was a

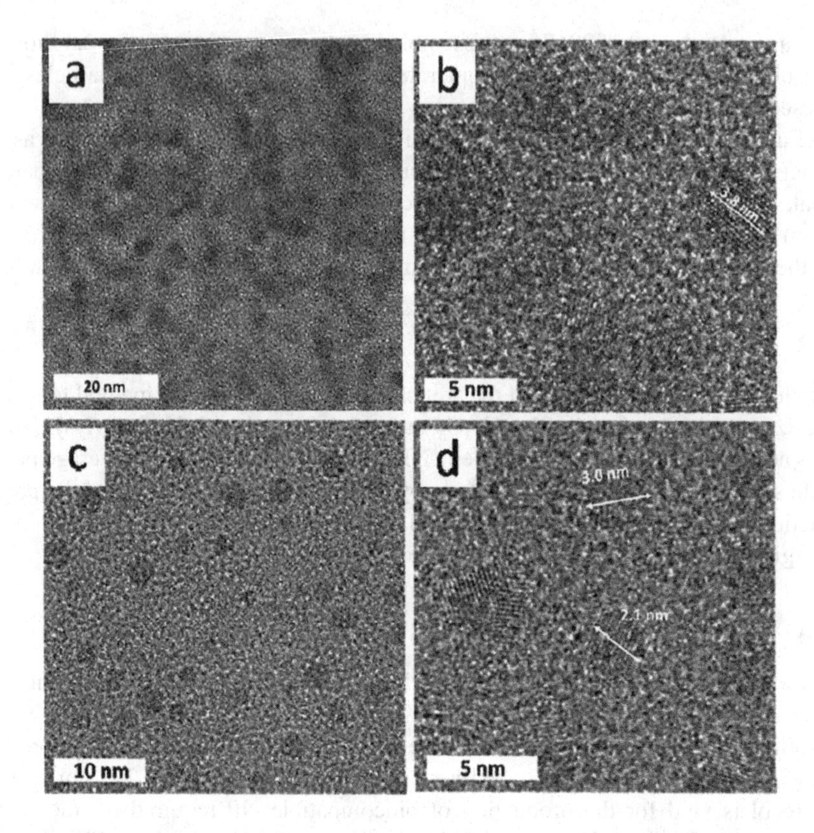

FIGURE 2.4 TEM images of the Fe NP obtained from hemoglobin ((a), (b)) and myoglobin ((c), (d)). The mean diameter of iron nanoparticles from hemoglobin and myoglobin is 3.2 nm and 3.6 nm, respectively [Reproduced with permission from Sayyad et al. (2012) © IOP Publishing].

combination of a slightly agglomerated sphere, rod, and dendritic nanostructure. Superparamagnetic characteristics were exhibited by Fe_3O_4 NPs prepared with galactose, mannose, and maltose. Consequently, the saccharides capped magnetite (Fe_3O_4) NPs can be used in biomedical imaging applications.

2.6 SYNTHETIC TANNIC AND GALLIC ACID

Both tannic and gallic acids are polyphenols and are considered as weak acids. They can be obtained from both natural and synthetic sources. Herrera-Becerra et al. (2010) obtained well-dispersed spherical iron oxide (Fe_2O_3) nanoparticles of less than 10 nm diameter with high crystallinity, using the synthetic tannic acid solution at pH10 combined with ultra-sonication. Similarly, Dorniani et al. (2012) prepared magnetic Fe NPs via the sonochemical process and applied a coat of Chitosan and Gallic acid on the nanoparticles to form a core-shell arrangement. XRD analysis confirmed that pure Fe_3O_4 nanoparticles were synthesized having

FIGURE 2.5 SEM images of samples. (a) the fresh carbon-encapsulated iron nanoparticles; (b) the carbon-encapsulated iron nanoparticles calcined under a nitrogen flow at 700°C; (c) the carbon-encapsulated iron nanoparticles after catalytic reaction [Reproduced with permission from Yan et al. (2015) © Elsevier].

cubic inverse spinel structure. The average particle size obtained was 11 nm with spherical structure. The synthesized magnetic NPs were found to be smaller than iron oxide-chitosan-gallic acid (FCG) nanostructure, having an average diameter of 13 nm. The prepared nanoparticles showed magnetic saturation of 26.07 emu/g. The FCG NPs were demonstrated to be capable of targeted drug delivery.

2.7 POLYSACCHARIDES

Polysaccharides are polymeric carbohydrates with glycosidic linkages. Chang et al. (2011) synthesized superparamagnetic Fe_3O_4 nanoparticles using different polysaccharides, like soluble starch, carboxymethyl cellulose sodium (CMC), and agar. Figure 2.6 shows the TEM images of the various coated Fe_3O_4 NPs synthesized by the research group.

The polysaccharides enhanced the biocompatibility and biodegradability of Fe_3O_4 NPs apart from acting as a capping and stabilizing agent. Starch stabilized nanoparticles were found to be 10 nm smaller than CMC and agar. The saturation magnetization of agar-based NPs was found to be 20.43 emu/g, whereas that of starch and CMC based NPS were found to be 36.16 emu/g, and 35.75 emu/g

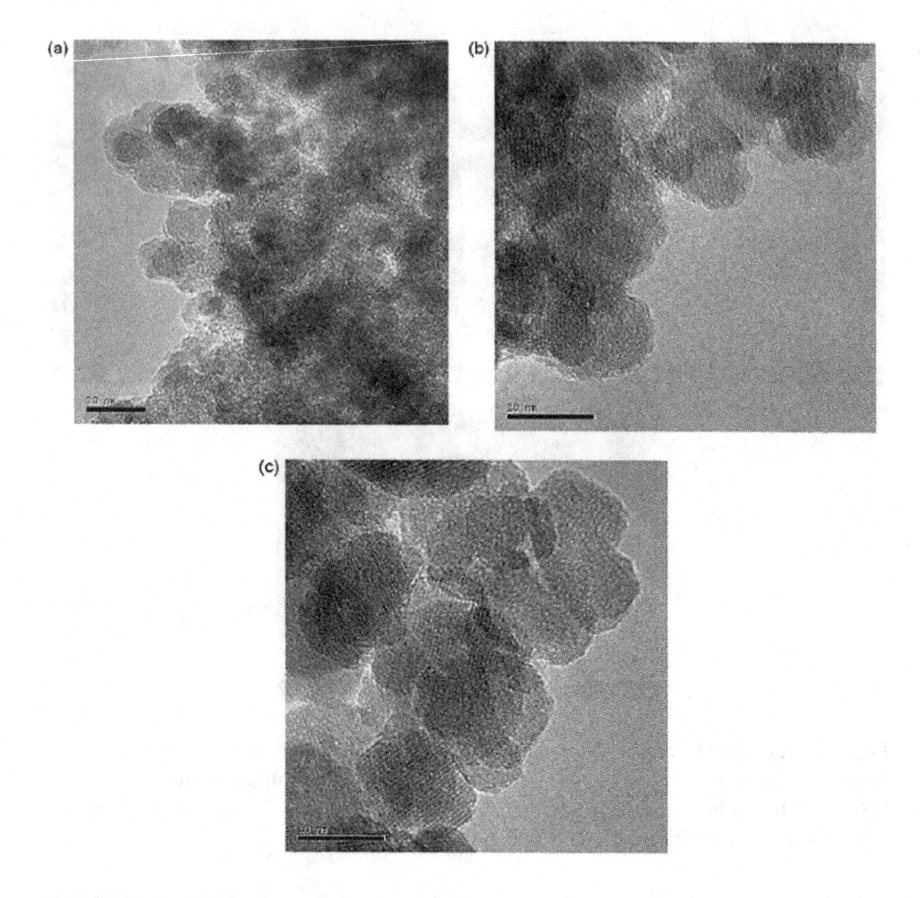

FIGURE 2.6 TEM micrographs for (a) soluble starch-coated Fe_3O_4 (b) carboxymethyl cellulose-coated Fe_3O_4 (b) and agar-coated Fe_3O_4 [Reproduced with permission from Chang et al. (2011) © Elsevier].

respectively. Extremely small hysteresis loop and coercivity characteristics were exhibited by the polysaccharide–Fe_3O_4 NPs.

2.8 CELLULOSE

Cellulose is the most abundant naturally occurring polymer in the world as it is an essential structural component of the primary cell wall of all plants, various types of algae, and the oomycetes (Klemm et al., 2005). Materials with high cellulosic content can be used to prepare and stabilize Fe NPs. López-Téllez et al. (2013) extracted cellulosic components from ethanol-cured powdered orange peel to prepare iron oxide nanorods. The study demonstrated the interaction between reduced metal ions and functional groups of cellulosic components through electrostatic and weak van der Waals forces. Such interactions helped stabilizing the nanoparticles after formation. The resultant nanorods acted like an adsorbent with an adsorption capacity of 7.44 mg/g for the removal of hexavalent chromium.

2.9 CLAY

Clay is fine particles of withered rocks, mainly consisting of phyllosilicate minerals, organic matter, and variable amounts of water trapped in the mineral structure. Clay is intensively used in making pottery, bricks and tiles. It has been suggested that clay can be utilized as supporting material for nanoparticle synthesis. Kalantari et al. (2014) used montmorillonite (MMT) as a rigid support for the synthesis of magnetite Fe_3O_4 nanoparticles. The Fe_3O_4 NP formation occurred in the interlayer space or on the surface of MMT, as confirmed with the help of TEM images. The saturation magnetization was found to increase from 12 to 32.4 emu/g for the MMT/Fe_3O_4 composite with Fe_3O_4 content increasing from 1 to 12 wt%. In another study, Ding et al. (2016) used a type of nanoclay smectite known as LAPONITE® to prepare LAPONITE®-Fe_3O_4 nanoparticle (LAP-Fe_3O_4-NPs) via coprecipitation method for in vivo magnetic resonance imaging of tumors. The LAP-Fe_3O_4-NPs is reported to have excellent colloidal stability (Figure 2.7).

In Table 2.1 various Fe NPs syntheses utilizing numerous biocompatible green reagents acting as reducing and capping agents are mentioned.

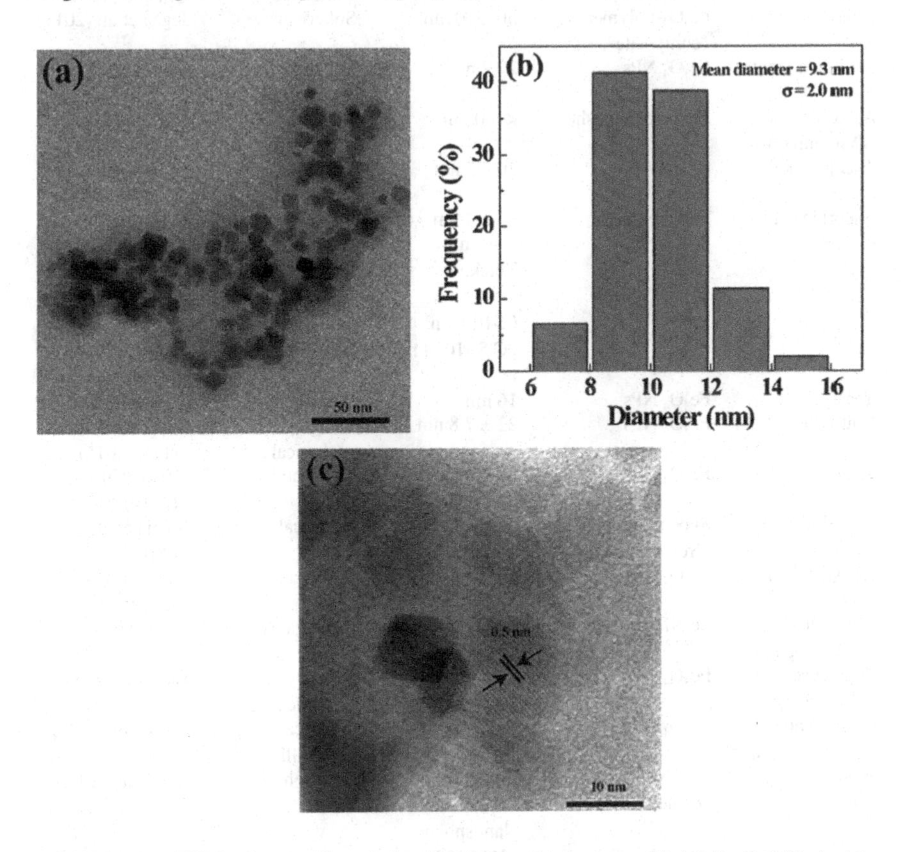

FIGURE 2.7 TEM micrograph and size distribution histogram of LAP-Fe_3O_4 NPs (a, b). (c) The high magnification TEM image of LAP-Fe_3O_4 NPs [Reproduced with permission from Ding et al. (2016) © Royal Society of Chemistry].

TABLE 2.1
Iron based Nanoparticle Synthesis Utilizing Biocompatible Green Reagents

Green Reagents	Nanoparticle	NPs Size	Morphology	Reference
α-D-glucose	Fe_3O_4 NPs	~12.5 nm	Spherical	Lu et al. (2010)
Starch	Fe/Pd bimetallic nanoparticle	14.1 nm	Discrete (well dispersed)	He and Zhao (2005)
Maltose	Fe_3O_4 NPs	12.1 ± 2.1 nm	Spherical	Demir et al. (2013)
Sucrose	Fe_3O_4 NPs	4–16 nm	Spherical	Sun et al. (2009)
α-D-maltose	Fe_3O_4 NPs	9.7 ± 1.0 nm	Mostly spherical	Demir et al. (2014)
α-D-mannose	Fe_3O_4 NPs	13.1 ± 0.3 nm	Rod-like	
α-D-galactose	Fe_3O_4 NPs	12.4 ± 0.3 nm	Spherical	
α-D-lactose	Fe_3O_4 NPs	3.8 ± 0.21 nm	Dendritic nanostructure	
D-glucose	Fe_3O_4 NPs	10–20 nm	Coral-like	Qin et al. (2011)
Agar	Fe_3O_4-polymer composite	50–200 nm	Spherical	Jegan et al. (2011)
Sodium alginate	Fe_3O_4 NPs	27.2 nm	Spherical, and hexagonal	Gao et al. (2008)
Ascorbic acid (Vitamin C)	Fe metal nanoshell	<100 nm	Cubical	Nadagouda and Varma (2007)
Ascorbic acid	nZVI	20–75 nm	Spherical chain	Savasari et al. (2015)
Ascorbic acid	Fe_3O_4 NPs	15 ± 4 nm	Irregular	Nene et al. (2016)
Urease	Fe_3O_4 NPs	19±5 nm Thickness <2–3 nm L>100 nm, C.S~10±4 nm	Nanosphere at 60°C Nanosheet at 40°C Nanorods at 40°C	Shi et al. (2014)
Yeast	Fe_3O_4 NPs	16 nm	Wormhole-like	Zhou et al. (2009)
Chitosan	Fe_3O_4 NPs	22 ± 7.8 nm	Roughly spherical	Shrifian-Esfahni et al. (2015)
Agar	Fe_3O_4 NPs	20–30 nm	Nonspherical	Hsieh et al. (2010)
Ascorbic acid	Superparamagnetic iron oxide	5 nm	Spherical	Sreeja et al. (2014)
L-lysine amino acid	Fe_3O_4 NPs	17.5 nm	Spherical	Krishna et al. (2012)
Hemoglobin and myoglobin	Fe NPs	2–5 nm	Aggregates	Sayyad et al. (2012)
D-glucose gluconic acid	Fe_3O_4 NPs	12.5 nm	Roughly spherical	Lu et al. (2010)
Glucose and gluconic acid	Fe_3O_4 NPs	4–16 nm	Spherical crystalline	Sun et al. (2009)
Wood derived sugar	Iron NPs carbon coated	Iron core: 10–25 nm Nanosphere: 100–150 nm	Nanosphere	Yan et al. (2015)
Tannic acid	Iron Oxide	<10 nm	Spherical	Herrera-Becerra et al. (2010)

(Continued)

TABLE 2.1 (Continued)

Green Reagents	Nanoparticle	NPs Size	Morphology	Reference
Chitosan-Gallic acid	Fe NPs core shell	~11 nm	Cubical	Dorniani et al. (2012)
Starch	Fe_3O_4 NPs	<10 nm	Spherical	Chang et al. (2011)
Carboxymethyl cellulose sodium	Fe_3O_4 NPs	>10 nm	Spherical	Chang et al. (2011)
Agar	Fe_3O_4 NPs	>10 nm	Spherical	Chang et al. (2011)
Pectin	Fe_3O_4 NPs	5–18 nm	Cubical	Namanga et al. (2013)
Arginine	Fe_3O_4 NPs	Fe/Ar (1:1)-5-18 nm Fe/Ar (1:2)-9-15 nm	Spherical	Wang et al. (2009)

Source: [Reproduced with permission from Mondal et al. (2020) © Elsevier].

REFERENCES

Chang, P.R., Yu, J., Ma, X., Anderson, D.P. 2011. Polysaccharides as stabilizers for the synthesis of magnetic nanoparticles. *Carbohydr. Polym.* 83 (2), 640–644.

Demir, A., Topkaya, R., Baykal, A. 2013. Green synthesis of superparamagnetic Fe_3O_4 nanoparticles with maltose: Its magnetic investigation. *Polyhedron* 65, 282–287.

Demir, A., Baykal, A., Sözeri, H., 2014. Green synthesis of Fe_3O_4 nanoparticles by one-pot saccharide-assisted hydrothermal method. *Turk. J. Chem.* 38, 825–836.

Ding, L., Hu, Y., Luo, Y., Zhu, J., Wu, Y., Yu, Z., Cao, X., Peng, C., Shi, X., Guo, R. 2016. LAPONITE®-stabilized iron oxide nanoparticles for in vivo MR imaging of tumors. *Biomat. Sci.* 4 (3), 474–482.

Dorniani, D., Hussein, M.Z., Kura, A.U., Fakurazi, S., Shaari, A.H., Ahmad, Z. 2012. Preparation of Fe_3O_4 magnetic nanoparticles coated with gallic acid for drug delivery. *Int. J. Nanomed.* 7, 5745–5756.

Gao, S., Shi, Y., Zhang, S., Jiang, K., Yang, S., Li, Z., Takayama-Muromachi, E. 2008. Biopolymer-assisted green synthesis of iron oxide nanoparticles and their magnetic properties. *J. Phys. Chem. C* 112, 10398–10401.

He, F., Zhao, D. 2005. Preparation and characterization of a new class of starch-stabilized bimetallic nanoparticles for degradation of chlorinated hydrocarbons in water. *Environ. Sci. Technol.* 39, 3314–3320.

Herrera-Becerra, R., Rius, J.L., Zorrilla, C. 2010. Tannin biosynthesis of iron oxide nanoparticles. *Appl. Phys. A* 100, 453–459.

Hsieh, S., Huang, B., Hsieh, S., Wu, C., Wu, C., Lin, P., Huang, Y., Chang, C. 2010. Green fabrication of agar-conjugated Fe_3O_4 magnetic nanoparticles. *Nanotechnology* 21, 1–6.

Jegan, A., Ramasubbu, A., Saravanan, S., Vasanthkumar, S. 2011. One-pot synthesis and characterization of biopolymer—Iron oxide nanocomposite. *Int. J. Nano Dimens.* 2, 105–110.

Kalantari, K., Ahmad, M.B., Shameli, K., Hussein, M.Z.B., Khandanlou, R., Khanehzaei, H., 2014. Size-controlled synthesis of Fe_3O_4 magnetic nanoparticles in the layers of montmorillonite. *J. Nanomat.* https://doi.org/10.1155/2014/739485.

Klemm, D., Heublein, B., Fink, H.P., Bohn, A. 2005. Cellulose: Fascinating biopolymer and sustainable raw material. *Angew. Chem. Int. Ed.* 44 (22), 3358–3393.

Krishna, R., Titus, E., Krishna, R., Bardhan, N., Bahadur, D., Gracio, J. 2012. Wet-chemical green synthesis of L-lysine amino acid stabilized biocompatible iron-oxide magnetic nanoparticles. *J. Nanosci. Nanotechnol.* 12, 6645–6651.

López-Téllez, G., Balderas-Hernández, P., Barrera-Díaz, C.E., Vilchis-Nestor, A.R., Roa-Morales, G., Bilyeu, B. 2013. Green method to form iron oxide nanorods in orange peels for chromium (VI) reduction. *J. Nanosci. Nanotech.* 13 (3), 2354–2361.

Lu, W., Shen, Y., Xie, A., Zhang, W. 2010. Green synthesis and characterization of superparamagnetic Fe_3O_4 nanoparticles. *J. Magn. Magn. Mater.* 322, 1828–1833.

Mondal, P., Anweshan, A., Purkait, M.K. 2020. Green synthesis and environmental application of iron-based nanomaterials and nanocomposite: A review. *Chemosphere* 259, 127509.

Nadagouda, M.N., Varma, R.S. 2007. A greener synthesis of core (Fe, Cu)-shell (Au, Pt, Pd, and Ag) nanocrystals using aqueous vitamin C. *Cryst. Growth Des.* 7 (12), 2582–2587.

Namanga, J., Foba, J., Ndinteh, D.T., Yufanyi, D.M., Krause, R.W.M. 2013. Synthesis and magnetic properties of a superparamagnetic nanocomposite pectin-magnetite nanocomposite. *J. Nanomater.* 2013, 1–8.

Nene, A.G., Takahashi, M., Wakita, K., Umeno, M. 2016. Size controlled synthesis of Fe_3O_4 nanoparticles by ascorbic acid mediated reduction of $Fe(acac)_3$ without using capping agent. *J. Nano. Res.* 40, 8–19.

Qin, Z., Jiao, X., Chen, D. 2011. Preparation of coral-like magnetite through a glucose assisted solvo-thermal synthesis. *Cryst. Eng. Comm.* 13, 4646–4651.

Savasari, M., Emadi, M., Bahmanyar, M.A., Biparva, P. 2015. Optimization of Cd(II) removal from aqueous solution by ascorbic acid-stabilized zero valent iron nanoparticles using response surface methodology. *J. Ind. Eng. Chem.* 21, 1403–1409.

Sayyad, A.S., Balakrishnan, K., Ci, L., Kabbani, A.T., Vajtai, R., Ajayan, P.M. 2012. Synthesis of iron nanoparticles from hemoglobin and myoglobin. *Nanotechnology* 23, 055602.

Shi, H., Tan, L., Du, Q., Chen, X., Li, L., Liu, T., Fu, C., Liu, H., Meng, X. 2014. Green synthesis of Fe_3O_4 nanoparticles with controlled morphologies using urease and their application in dye adsorption. *Dalton Trans.* 43, 12474–12479.

Shrifian-Esfahni, A., Salehi, M.T., Nasr-Esfahni, M., Ekramian, E. 2015. Chitosan-modified superparamgnetic iron oxide nanoparticles: design, fabrication, characterization and antibacterial activity. *Chemik* 69, 19–32.

Siskova, K.M., Straska, J., Krizek, M., Tucek, J., Machala, L., Zboril, R. 2013. Formation of zero-valent iron nanoparticles mediated by amino acids. *Proc. Environ. Sci.* 18, 809–817.

Sreeja, V., Jayaprabha, K.N., Joy, P.A. 2014. Water-dispersible ascorbic-acid-coated magnetite nanoparticles for contrast enhancement in MRI. *Appl. Nanosci.* 5, 435–441.

Sun, X., Zheng, C., Zhang, F., Yang, Y., Wu, G., Yu, A., Guan, N. 2009. Size-controlled synthesis of magnetite (Fe_3O_4) nanoparticles coated with glucose and gluconic acid from a single Fe(III) precursor by a sucrose bifunctional hydrothermal method. *J. Phys. Chem. C* 113, 16002–16008.

Wang, Z., Zhu, H., Wang, X., Yang, F., Yang, X. 2009. One-pot green synthesis of biocompatible arginine-stabilized magnetic nanoparticles. *Nanotechnology* 20, 1–10.

Yan, Q., Street, J., Yu, F. 2015. Synthesis of carbon-encapsulated iron nanoparticles from wood derived sugars by hydrothermal carbonization (HTC) and their application to convert bio-syngas into liquid hydrocarbons. *Bio. Bioenergy*. 83, 85–95.

Zhou, W., He, W., Zhong, S., Wang, Y., Zhao, H., Li, Z., Yan, S., 2009. Biosynthesis and magnetic properties of mesoporous Fe_3O_4 composites. *J. Magn. Magn. Mater*. 321, 1025–1028.

3 Microorganism-based Synthesis

3.1 INTRODUCTION

The synthesis of nanoparticles based on microorganisms has gained substantial popularity due to its benefits compared with traditional chemical protocols. The advantages are listed as: synthesis at ambient conditions, high energy-efficiency, limited production of toxic by-products, naturally abundant and renewable precursors, robust and simple scaling-up (Park et al., 2016). Through extracellular or intracellular processes, microorganisms like fungi, bacteria, and yeast can synthesize nanoparticles.

These mechanisms involve reduction of metal ions by enzymes along with producing well-dispersed nanoparticles with lower average particle size distribution. The nanoparticle surface gets incorporated with natural proteins, tannins, and peptides as capping agents. Such surface coating provides better stability and dispersion of nanoparticles by reducing agglomeration (Singh et al., 2016). The intracellular mechanism involves diffusion of metal ions into the cell, and the enzymes present reduce the ions to form nanoparticles. Whereas the extracellular mechanism involves electrostatic attraction of metal ions to the cell wall and carrying out the enzymatic reduction of metal ions. Table 3.1 represents various microorganisms utilized by various research groups for synthesizing iron and iron-based nanomaterials as reducing and capping agents.

3.2 BACTERIA UTILIZED SYNTHESIS

In their study Bharde et al. (2008) reported the use of *Actinobacter* sp. under aerobic conditions to prepare spherical iron oxide nanoparticles. In another study, the same bacterium species was used to synthesize maghemite (γ-Fe_2O_3) and greigite (Fe_3S_4) by changing the metal ion precursor (Bharde et al., 2005). Furthermore, it was established that *Actinobacter* sp. employed extracellular mechanism, involving iron reductase enzyme, to synthesize the magnetic nanoparticles. Figure 3.1 shows the TEM images of as-synthesized and calcined maghemite nanoparticles.

Moon et al. (2010) in a similar study synthesized mono-dispersed, sphere-shaped 13.1 nm magnetite Fe_3O_4 NPs under anaerobic conditions from FeOOH using *Thermoanaerobacter* sp., an extremophile. The results suggest that around $500/kg is the estimated cost for commercial production of nanomagnetite (25–50 nm).

In comparison to chemical synthesis, a fraction of cost is required to produce 5–90 nm pure or substituted magnetites through microbial process. Sundaram

DOI: 10.1201/9781003243632-3

TABLE 3.1
Microorganism Mediated Synthesis of Iron-based Nanoparticles

Microorganisms	Species Name	Nanoparticle	Average Size	Morphology	Reference
Bacteria	*Actinobacter sp.*	Fe_3O_4	10–40 nm	Cubical	Bharde et al. (2005)
	Actinobacter sp.	γ-Fe_2O_3	<50 nm	Spherical	Bharde et al. (2008)
	Thermoanaerobacter sp.	Fe_3O_4	~13 nm	Spherical	Moon et al. (2010)
	Bacillus subtilis	Fe_3O_4	60–80 nm	Spherical	Sundaram et al. (2012)
	Thiobacillus thioparus	Fe_3O_4	–	–	Elcey et al. (2014)
	Microbacterium marinilacus	Magnetic Iron oxide nanoparticles	2–10 nm	Spherical	Mehrotra et al. (2017)
	Desulfovibrio, strain LS4	Maghemite (Fe_2O_3) nanoparticles	19 nm	Round shaped	Das et al. (2018)
Fungi	*Fusarium oxysporum and Verticillium sp.*	Fe_3O_4	20–50 nm	Spherical	Bharde et al. (2006)
	P. chlamydosporium, A. fumigates, A. wentii, C. lunata and C. globosum	Maghemite (Fe_2O_3) nanoparticles	~12–50 nm	Spherical	Kaul et al. (2012)
	Aspergillus sp.	Fe_3O_4	50–200 nm	Spherical	Pavani and Kumar (2013)
	Alternaria alternate	Fe Nanoparticles	~9nm	Cubical	Mohamed (2015)
	Pleurotus sp.	Fe Nanoparticles			Mazumdar and Haloi (2011)
	Aspergillus niger YESM 1	Magnetic Fe and Fe_3O_4 (magnetite) nanoparticles	18 and 50 nm for Fe and Fe_3O_4 NPs respectively	Spherical	Abdeen et al. (2016)
	Alternaria alternate	γ-Fe_2O_3 (iron oxide) nanoparticles	75–650 nm	Quasi-spherical as well as rectangular NPs	Sarkar et al. (2017)
Algae	*Sargassum muticum*	Magnetic Fe_3O_4 nanoparticles	18 ± 4 nm	Cubical	Mahdavi et al. (2013)
	Chlorococcum sp.	Iron nanoparticles	20–50 nm	Spherical	Subramaniyam et al. (2015)

Source: [Reproduced with permission from Mondal et al. (2020) © Elsevier].

et al. (2012) used *Bacillus subtilis* strains isolated from rhizosphere soil to prepare Fe_3O_4 nanoparticles with spherical morphology and 60–80 nm diameters. Figure 3.2 shows the SEM image of Fe_3O_4 nanoparticles obtained utilizing

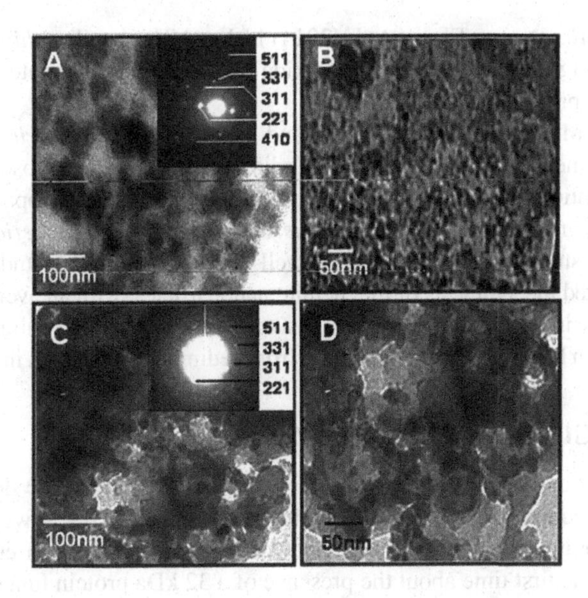

FIGURE 3.1 TEM images of as-synthesized (a and b) and calcined (c and d) maghemite nanoparticles. The insets in panels (a) and (c) show the SAED pattern of the as-synthesized and calcined maghemite nanoparticles respectively [Reproduced with permission from Bharde et al. (2008) © Elsevier].

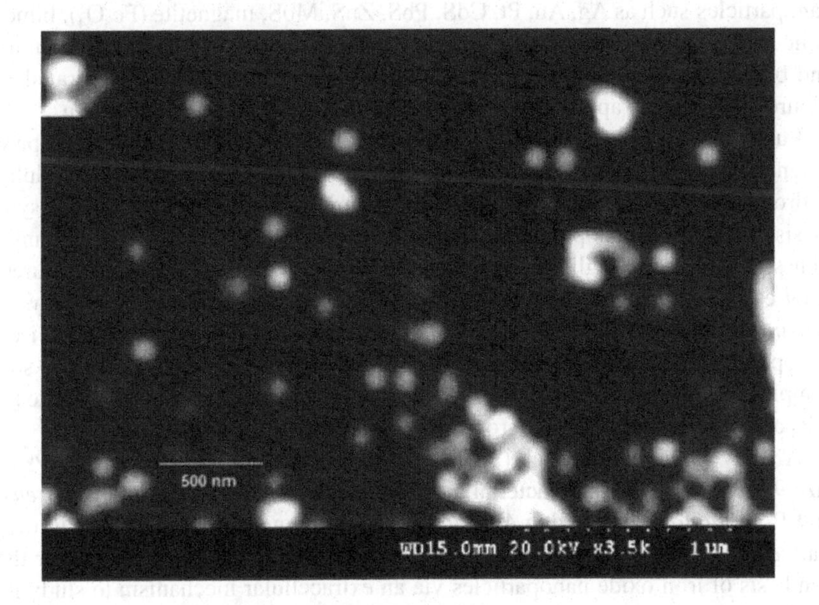

FIGURE 3.2 SEM image of Fe_3O_4 nanoparticles obtained utilizing *Bacillus subtilis* strains isolated from rhizosphere soil [Reproduced with permission from Sundaram et al. (2012) © Springer].

Bacillus subtilis strains. Elcey et al. (2014) isolated *Thiobacillus thioparus* from iron ore mining zones to produce iron oxide NPs termed as magnetosomes, having magnetic properties.

Likewise, Mukherjee (2017) isolated and incubated *Microbacterium marinilacus* from sediments of Damodar River in India to synthesize magnetosomes having a size distribution from 32 to 48 nm. Further, the synthesized nanoparticles were tested against gram positive *Bacillus cereus*, and gram negative *Escherichia coli* for antimicrobial studies, through agar-plate-well-diffusion. In another study, Das et al. (2018) reported the synthesis of maghemite nanoparticles, with an average diameter of 18 nm, in an anaerobic process, using native hypersaline sulfate-reducing bacterial strain LS4, isolated and cultured from sediment of a saltpan in Goa, India.

3.3 FUNGI MEDIATED SYNTHESIS

Functional groups such as ≡C–O–C≡, =C=O, ≡C–O–R and =C=C= derived from heterocyclic compounds found in proteins from fungal extracts were reported to act as capping ligands in NPs (Sanghi and Verma, 2009). Jain et al. (2011) reported for the first time about the presence of a 32 kDa protein functioning as a capping agent in Ag nanoparticles synthesized by *Aspergillus flavus* NJP08. The biosynthesis may be intracellular or extracellular or both. An added advantage is the presence of mycelia which provides a large surface area, leading to easy scale-up of nanoparticles on an industrial scale. Among several fungi, *Fusarium oxysporum* has been predominantly explored for the synthesis of a wide range of nanoparticles such as Ag, Au, Pt, CdS, PbS, ZnS, MoS, magnetite (Fe_3O_4), bimetallic Au-Ag alloy, silica, titanium, zirconia, quantum dots, magnetite, strontianite and barium titanate by various intra and extracellular mechanisms illustrated in Figure 3.3, when an appropriate salt is added to the fungal growth medium.

Fungi like *Fusarium oxysporum* and *Verticillium sp.* are exploited to prepare magnetic nanoparticles of different sizes at ambient temperature via extracellular hydrolysis of the anionic iron complexes. Bharde et al. (2006) reported the synthesis of magnetite nanoparticles using fungi that showed the ferrimagnetic transition signature with negligible spontaneous magnetization at low temperatures. Kaul et al. (2012) explored five different species of fungi, namely *P. chlamydosporium*, *A. fumigates*, *A. wentii*, *C. lunata*, and *C. globosum*, to prepare iron nanoparticles. Pavani and Kumar (2013) isolated *Aspergillus sp.* from the soil sample collected near metal plating industry in Hyderabad, India, to synthesize Fe NPs and consequently remove iron from wastewater.

A study by Mohamed et al. (2015) reported the formation of cubic Fe NPs of size 9 ± 3 nm, having antibacterial activity against *B. subtilis*, *E. coli*, *S. aureus*, and *P. aeruginosa*, employing *Alternaria* alternata fungus in lightless conditions. Sarkar et al. (2017) used *Alternaria alternate*, a phytopathogenic fungus for the synthesis of iron oxide nanoparticles via an extracellular mechanism to study its mechanical properties. Abdeen et al. (2016) explored green processes to synthesize magnetic Fe NPs and magnetite nanoparticles using *Asperigillus niger* via intracellular mechanism. Figure 3.4 shows the various methods for synthesis of magnetic nanoparticles.

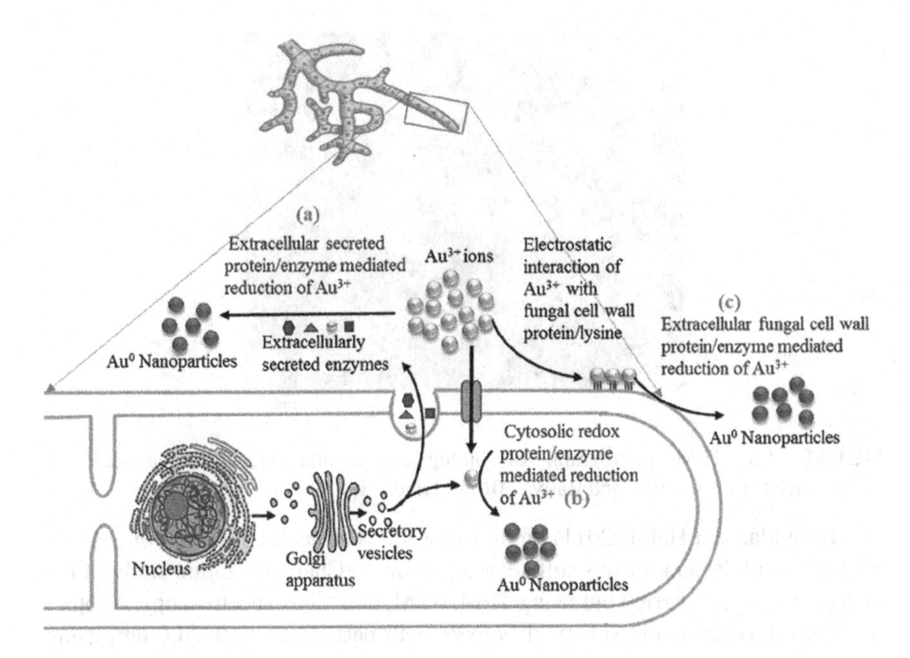

FIGURE 3.3 Different mechanisms for metal nanoparticle synthesis by fungi: (a) Gold (III)-ion is reduced into elemental gold by extracellularly secreted enzymes/proteins. (b) Cytosolic redox protein/enzyme mediated reduction of gold (III)-ion. (c) Gold-ion makes an electrostatic interaction with fungal cell wall protein/enzyme where it gets reduced extracellularly by cell wall protein/enzymes [Reproduced with permission from Rana et al. 2020 © Elsevier].

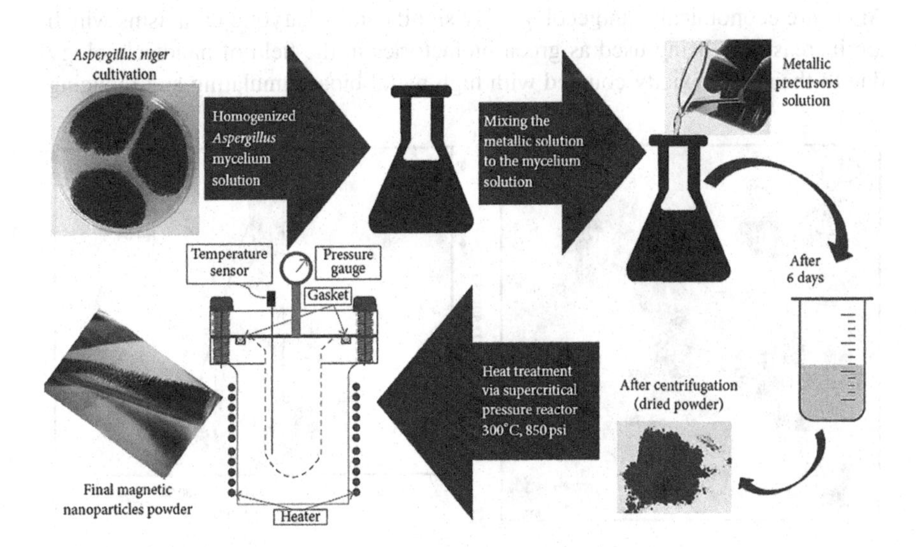

FIGURE 3.4 Schematic diagram of the biological-physical method for synthesis of magnetic nanoparticles [Reproduced from Abdeen et al. 2016 © Hindawi].

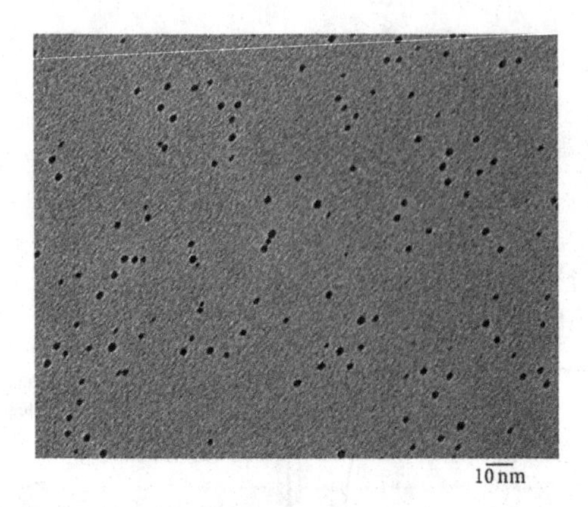

FIGURE 3.5 TEM micrograph of biologically synthesized iron nanoparticles [Reproduced from Tarafdar and Raliya 2013 © Hindawi].

Mazumdar and Haloi (2011) employed a filamentous fungus *Pleurotus sp.* to prepare iron NPs and ferrous sulfate as a precursor. Characterization study of the nanoparticles was carried out using TEM, SEM, and FTIR spectroscopy. Another method for synthesizing spherical Fe NPs with particle size 10–24.6 nm, using *Aspergillus oryzae* TFR9 (fungus), was proposed by Tarafdar and Raliya (2013) and found large application in the fields of engineering, agriculture, and biomedicine. Figure 3.5 depicts the TEM micrograph of biologically synthesized iron nanoparticles.

3.4 ALGAE

Algae are economically and ecologically significant eukaryotic organisms which are increasingly being used as green biofactories in the field of nanotechnology due to their low toxicity coupled with high metal bioaccumulating and reducing

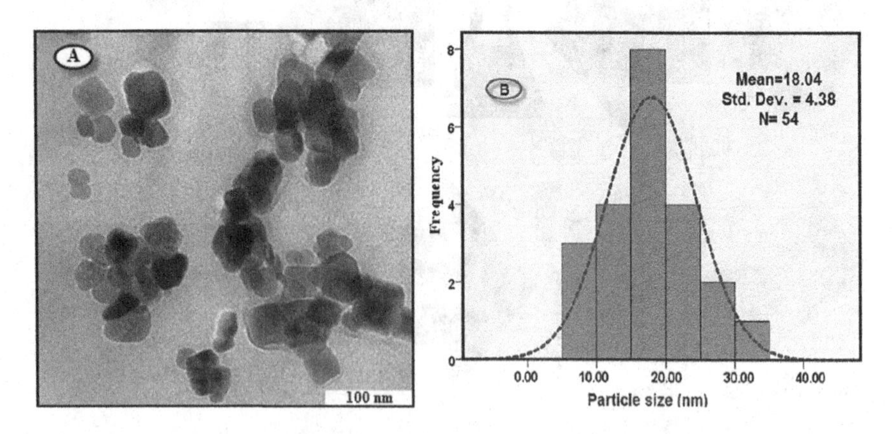

FIGURE 3.6 TEM image (a), and corresponding size distribution of Fe_3O_4-NPs synthesized using brown seaweed extract (b) [Reproduced from Mahdavi et al. (2013) © MDPI].

capabilities. Algae refer to a diverse group of photosynthetic eukaryotic organisms that are polyphyletic. They consist of both unicellular and multicellular organisms.

Mahdavi et al. (2013) used an extract from the macroalgae, brown seaweed (*Sargassum muticum*) to synthesize iron oxide (Fe_3O_4) nanoparticles as shown in Figure 3.6, with superparamagnetic characteristics and having an average particle size of 18 ± 4 nm. X-ray diffraction (XRD) revealed that the nanoparticles were crystalline and had a cubic morphology. The authors also reported that the functional bioactivity, i.e., antimicrobial activity of Fe_3O_4 NPs, prepared using the green synthesis route is comparably higher than the nanoparticles synthesized by conventional chemical methods.

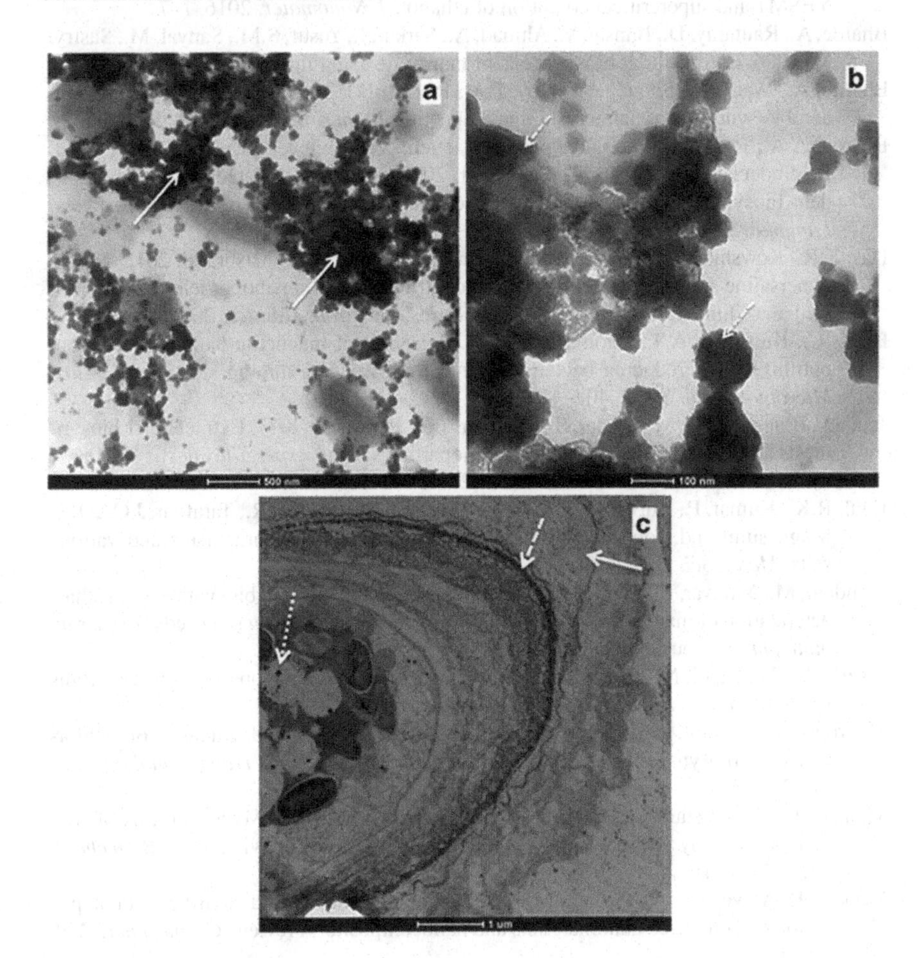

FIGURE 3.7 TEM images of phycosynthesized iron nanoparticle; (a) presence of nanoiron outside the cell (*solid arrows*), (b) capping of nanoiron by biomolecules (*dashed arrows*), (c) localization of nanoiron on the cell surface (*dashed arrows*), inside (*dashed arrows*) and outside (*solid arrows*) the cell [Reproduced with permission from Subramaniyam et al. (2015) © Springer].

A different study carried out by Subramaniyam et al. (2015) reported the formation of spherical-shaped Fe NPs as shown in Figure 3.7, ranging in size from 20 to 50 nm using soil microalgae, *Chlorococcum sp.*, with an iron chloride precursor. The study suggested that the amine and carbonyl functional groups on polysaccharides and glycoproteins present in the algal cell were involved in the nanoparticle formation and verified by FTIR analysis.

REFERENCES

Abdeen, M., Sabry, S., Ghozlan, H., El-Gendy, A.A., Carpenter, E.E. 2016. Microbial physical synthesis of Fe and Fe_3O_4 magnetic nanoparticles using *Aspergillus niger* YESM1 and supercritical condition of ethanol. *J. Nanomater.* 2016, 1–7.

Bharde, A., Rautaray, D., Bansal, V., Ahmad, A., Sarkar, I., Yusuf, S.M., Sanyal, M., Sastry, M. 2006. Extracellular biosynthesis of magnetite using fungi. *Small* 2, 135–141.

Bharde, A., Wani, A., Shouche, Y., Joy, P.A., Prasad, B.L.V., Sastry, M. 2005. Bacterial aerobic synthesis of nanocrystalline magnetite. *J. Am. Chem. Soc.* 127, 9326–9327.

Bharde, A.A., Parikh, R.Y., Baidakova, M., Jouen, S., Hannoyer, B., Enoki, T., Prasad, B., Shouche, Y.S., Ogale, S., Sastry, M. 2008. Bacteria-mediated precursor-dependent biosynthesis of superparamagnetic iron oxide and iron sulfide nanoparticles. *Langmuir* 24, 5787–5794.

Das, K.R., Kowshik, M., Kumar, M.P., Kerkar, S., Shyama, S.K., Mishra, S. 2018. Native hypersaline sulphate reducing bacteria contributes to iron nanoparticle formation in saltpan sediment: A concern for aquaculture. *J. Environ. Manage.* 206, 556–564.

Elcey, C., Kuruvilla, A.T., Thomas, D. 2014. Synthesis of magnetite nanoparticles from optimized iron reducing bacteria isolated from iron ore mining sites. *Int. J. Curr. Microbiol. Appl. Sci.* 3, 408–417.

Jain, N., Bhargava, A., Majumdar, S., Tarafdar, J.C., Panwar, J. 2011. Extracellular biosynthesis and characterization of silver nanoparticles using *Aspergillus flavus* NJP08: A mechanism prospective. *Nanoscale* 3 (2), 635–641.

Kaul, R.K., Kumar, P., Burman, U., Joshi, P., Agrawal, A., Raliya, R., Tarafdar, J.C. 2012. Magnesium and iron nanoparticles production using microorganisms and various salts. *Mater. Sci. Poland.* 30, 254–258.

Mahdavi, M., Namvar, F., Ahmad, M.B., Mohamad, R. 2013. Green biosynthesis and characterization of magnetic iron oxide (Fe_3O_4) nanoparticles using seaweed (*Sargassum muticum*) aqueous extract. *Molecules* 18 (5), 5954–5964.

Mazumdar, H., Haloi, N. 2011. A study on biosynthesis of iron nanoparticles by Pleurotus sp. *J. Microbiol. Biotech. Res.* 1 (3), 39–49.

Mehrotra, N., Tripathi, R.M., Zafar, F., Singh, M.P. 2017. Catalytic degradation of dichlorvos using biosynthesized zero valent Iron nanoparticles. *IEEE Trans. Nanobiosci.* 16 (4), 280–286.

Mohamed, Y.M., Azzam, A.M., Amin, B.H., Safwat, N.A. 2015. Mycosynthesis of iron nanoparticles by *Alternaria alternata* and its antibacterial activity. *Afr. J. Biotechnol.* 14, 1234–1241.

Mondal, P., Anweshan, A., Purkait, M.K. 2020. Green synthesis and environmental application of iron-based nanomaterials and nanocomposite: A review. *Chemosphere* 259, 127509.

Moon, J.W., Rawn, C.J., Rondinone, A.J., Love, L.J., Roh, Y., Everett, S.M., Lauf, R.J., Phelps, T.J. 2010. Large-scale production of magnetic nanoparticles using bacterial fermentation. *J. Ind. Micro. Biotech.* 37 (10), 1023–1031.

Mukherjee, P. 2017. Stenotrophomonas and microbacterium: Mediated biogenesis of copper, silver and iron nanoparticles—proteomic insights and antibacterial properties versus biofilm formation. *J. Cluster Sci.* 28 (1), 331–358.

Park, T.J., Lee, K.G., Lee, S.Y., 2016. Advances in microbial biosynthesis of metal nanoparticles. *App. Micro. Biotechnol.* 100 (2), 521–534.

Pavani, K.V., Kumar, N.S. 2013. Adsorption of iron and synthesis of iron nanoparticles by Aspergillus species kvp 12. *Am. J. Nanomater.* 1, 24–26.

Rana, A., Yadav, K., Jagadevan, S. 2020. A comprehensive review on green synthesis of nature-inspired metal nanoparticles: Mechanism, application and toxicity. *J. Clean. Prod.* 272, 122880.

Sanghi, R., Verma, P. 2009. Biomimetic synthesis and characterisation of protein capped silver nanoparticles. *Bioresour. Technol.* 100, 501–504.

Sarkar, J., Mollick, M.M.R., Chattopadhyay, D., Acharya, K. 2017. An eco-friendly route of γ-Fe_2O_3 nanoparticles formation and investigation of the mechanical properties of the HPMC-γ-Fe_2O_3 nanocomposites. *Bioprocess Biosyst. Eng.* 40 (3), 351–359.

Singh, P., Kim, Y.J., Zhang, D., Yang, D.C. 2016. Biological synthesis of nanoparticles from plants and microorganisms. *Trends Biotechnol.* 34 (7), 588–599.

Subramaniyam, V., Subashchandrabose, S.R., Thavamani, P., Megharaj, M., Chen, Z., Naidu, R. 2015. Chlorococcum sp. MM11-A novel phyco-nanofactory for the synthesis of iron nanoparticles. *J. Appl. Phycol.* 27, 1861–1869.

Sundaram, P.A., Augustine, R., Kannan, M. 2012. Extracellular biosynthesis of iron oxide nanoparticles by *Bacillus subtilis* strains isolated from rhizosphere soil. *Biotechnol. Bioprocess Eng.* 17, 835–840.

Tarafdar, J.C., Raliya, R. 2013. Rapid, low-cost, and ecofriendly approach for Iron nanoparticle synthesis using *Aspergillus oryzae* TFR9. *J. Nanopart.* 2013, 1–4.

4 Plant Source Mediated Synthesis

The nanoparticles prepared from microorganisms as discussed in Chapter 3, were found to have low dispersion, and the rate of formation was also slow when compared to plant-mediated synthesis (Dhillon et al., 2012). A study by Kalaiarasi et al. (2010) reported that plant-based synthesis of metallic nanoparticles is facile, most cost-efficient, and has high repeatability and reproducibility. It has been well established that plant extracts are most suitable for producing highly stable metal NPs at rapid rates and large quantities (Iravani, 2011). Plant-based nanomaterial synthesis is preferred as a plethora of different naturally occurring, and robust biomolecular reducing agents can be extracted from various plants (Mukunthan and Balaji, 2012). Various herbs and plants have a high concentration of antioxidants as active phytochemical constituents in fruits, leaves, seeds, and stems. The production of nanomaterials from processes utilizing botanical phytochemicals significantly reduces environmental pollution, thereby setting benchmarks in economically feasible and ultra-sustainable clean and green technologies (Zambre et al., 2012). Table 4.1 represents various plant parts utilized as a source for reducing agent and capping agents in synthesizing iron and iron-based nanomaterials by several researchers.

TABLE 4.1
Iron Nanoparticles Prepared from Various Plant Parts

Part	Name of Source	Average Size	Morphology	Reference
Plant	Soya bean sprouts	~8 nm	Spherical	Cai et al. (2010)
	Aloe vera	93–227 nm	Spherical	Ngernpimai et al. (2012)
	Aloe vera	~6–30 nm	Agglomerated irregular	Phumying et al. (2012)
	Alfalfa biomass	3.1 nm	Agglomerated	Herrera-Becerra et al. (2007)
	Pine wood shavings	1–10 nm	Irregular, rod and spherical shaped	Ramasahayam et al. (2012)
Marine plant	*Sargassum muticum*	18 ± 4 nm	Cubic	Mahdavi et al. (2013)
	Kappaphycus alvarezii	14.7 ± 1.8 nm	Spherical	Yew et al. (2016)
	Padina pavonica	10–19.5 nm	Spherical	El-Kassas Hala et al. (2016)
	Sargassum acinarium	21.6–27.4 nm	Spherical	El-Kassas Hala et al. (2016)

(*Continued*)

DOI: 10.1201/9781003243632-4

TABLE 4.1 (Continued)

Part	Name of Source	Average Size	Morphology	Reference
Seed	Grape seed proanthocyanidin	~30 nm	Irregular shape	Narayanan et al. (2011)
	Syzygium cumini	9–20 nm	Agglomerated spherical	Venkateswarlu et al. (2014)
	Carom and clove	0.088–3.95	Spherical, irregular	Afsheen et al. (2018)
Leaf	Carob	4–8 nm	Well monodispersed	Awwad and Salem (2012)
	Tridax procumbens	<100 nm	Irregular shape	Senthil and Ramesh (2012)
	Artemisia annua	3–10 nm	Spherical	Basavegowda et al. (2014a)
	Caricaya Papaya	33 nm	Agglomerated plate and capsule-like structures	Latha and Gowri (2014)
	Perilla frutescens	~50 nm	Spherical	Basavegowda et al. (2014b)
	Euphorbia wallichii	10–15 nm	Spherical	Atarod et al. (2015)
	Green tea	5.7 ± 4.1 nm	Spherical	Xiao et al. (2015)
	Zea mays L.	<100 nm	Aggregated spherical	Patra and Baek (2017)
	Sesbania grandiflora	25–60 nm	Agglomerated nonspherical	Rajendran and Sengodan (2017)
	Rubus glaucus Benth	40–70 nm	Aggregated spherical	Kumar et al. (2015)
	Calliandra haematocephala	~85.4–87.9 nm	Bead-like spherical	Sirdeshpande et al. (2018)
	Lagenaria siceraria	30–100 nm	Cubic	Kanagasubbulakshmi and Kadirvelu (2017)
	Moringa oleifera	~4.14 nm	Spherical and irregular	Silveira et al. (2017)
	Coriandrum sativum	20–90 nm	Spherical	Sathya et al. (2017)
	Eucalyptus	20–50 nm	Spherical and irregular	Weng et al. (2017)
	Eucalyptus	80–90 nm	Spherical	Gan et al. (2018)
	Eucalyptus	20–70 nm	Spherical	Sangami and Manu (2017)
	Oak	20–100 nm	Irregular shape	Machado et al. (2017)
	Lantana camara	10–20 nm	Crystalline nanorods	Rajiv et al. (2017)
	Cynometra ramiflora	>100 nm	Spherical aggregated	Groiss et al. (2017)
	Eichhornia crassipes	>100 nm	Rod shaped	Jagathesan and Rajiv (2018)
	Syzygium jambos (L.) Alston	13.7 ± 5.0 nm	Spherical	Xiao et al. (2017)
	Vaccinium corymbosum	52.4 nm	Irregular shape non agglomerated	Manquian-Cerda et al. (2017)

(*Continued*)

TABLE 4.1 (Continued)

Part	Name of Source	Average Size	Morphology	Reference
	Azadirachta indica (neem)	38 nm	Brick-like structure	Madhubala and Kalaivani (2017)
	Mangifera indica L.	3.0 ± 0.2 nm	Polycrystalline nanorods	Al-Ruqeishi et al. (2016)
	Mansoa alliacea (Garlic vine)	18.22 nm	Spherical nanoparticles	Prasad (2016)
	Eichhornia crassipes	20–80 nm	Amorphous NPs	Wei et al. (2017)
Fruit	*Passiflora tripartita*	18.2–24.7 nm	Spherical	Kumar et al. (2015)
	Averrhoa carambola	1.9–3.1 nm	Spherical	Ahmed et al. (2015)
	Lemon	14–17 nm	Spherical	Bahadur et al. (2017)
	Couroupita guianensis	17 ± 10 nm	Spherical	Jha (2017)
	Pomegranate	Agglomerate	100–200 nm	Rao et al. (2013)
	Pepper	50 nm	Dendrite-like	Khaghani and Ghanbari (2017)
	Terminalia chebula	<80 nm	Amorphous	Mohan Kumar et al. (2013)
	Zanthoxylum rhetsa	12.2 ± 0.8 nm	Cluster-like	Saikia et al. (2017)
Root & Shoot	*Mimosa pudica* (root)	60–80 nm	Agglomerated rough spherical	Niraimathee et al. (2016)
	Vaccinium corymbosum (shoot)	52.4 nm	Irregular shape, nonagglomerated	Manquian-Cerda et al. (2017)
Stolon	Potato	40 ± 2.2 nm	Cubic	Buazar et al. (2016)

Source: [Reproduced with permission from Mondal et al. (2020)].

4.1 SYNTHESIS BY LEAF EXTRACT

Extracts from the leaves of various plants have been extensively studied to estimate their polyphenolic contents along with other phytochemicals that have medicinal value. Xiao et al. (2017) studied the primary bioactive components of fifteen plant species and their capability to form zero-valent iron (ZVI) nanoparticles with an average size of 5 nm as shown in Figure 4.1. The study concluded that for preparation of Fe NPs polyphenol plays the most active role. Experiments concluded the other important biomolecules apart from polyphenols are reducing sugars, flavonoids and proteins. Extracts of *S. jambos (L.) Alston* (SJA) and *D. longan lour* was better at reducing Fe^{3+} than other plants, and the NPs formed thereafter had better capacity to remove chromium (VI) from solution. A maximum adsorption capacity of 983.2 mg Cr(VI)/g was obtained for (SJA) based iron NPs.

Optimization of several parameters, like precursor moiety, solvent, ration of biomass to the volume of solvent, amount of extract, pH, and temperature of the solution, needs to be carried out during nanoparticle synthesis. Huang et al. (2015) investigated the factors that determine the efficiency of Fe NPs prepared using

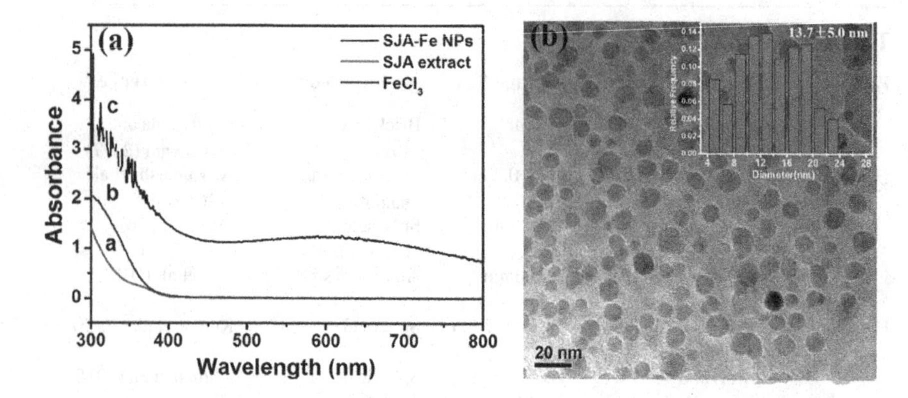

FIGURE 4.1 UV-Vis spectra of $FeCl_3$ (a–a), SJA extract (a–b) and reaction product (a–c) of $FeCl_3$ and *Syzygium jambos (L.) Alston* extract. TEM image of the synthesized Fe NPs (b) (inset indicates the size distribution) [Reproduced with permission from Xiao et al (2017 © Elsevier].

green tea leaf extracts for the removal of malachite green dye. The study explored the variation in the volume of tea extract, temperatures, and pH of the solution to understand the optimum conditions and reported that increasing the volume of extract and pH of the solution negatively impacted dye removal. However, increasing temperature improved nanoparticle reactivity. A maximum dye degradation of 90.56% was obtained by the prepared Fe NPs. Prasad (2016) synthesized β-Fe_2O_3 NPs using garlic vine leaf extract at pH 6.0 and iron (II) sulfate heptahydrate as a precursor. The author reported that the nanoparticle formed has a crystallite size of 18.22 nm with an energy band gap of 2.84 eV and hence, applicable as an indirect semiconductor.

Wang et al. (2014a) prepared spheroidal 20–80 nm, Fe NPs employing extract from dried eucalyptus leaf for the treatment of eutrophic wastewater. Through investigation it was observed about 71.7% of total N and 84.5% of COD was removed from the wastewater. Machado et al. (2013) conducted a study in which leaf extracts were prepared from twenty-six species of trees, namely apple, apricot, avocado, cherry, eucalyptus, kiwi, lemon, mandarin, medlar, mulberry, oak, olive, orange, passion fruit, peach, pear, pine, pomegranate, plum, quince, raspberry, strawberry, tea-black, tea-green, vine, and walnut to evaluate the viability of each extract capable of reducing trivalent iron in aqueous solution to form ZVI NPs of diameter 10–20 nm. The study established that dried leaves produce extracts with higher antioxidant content than fresh leaves, moreover, the most favorable extraction conditions varied for each leaf. The leaf extract obtained from oak, pomegranate, and green tea at 80°C performed the best of all.

In another study, Machado et al. (2013) explored the application of the green-synthesized ZVI NPs to remove one of the prevalent anti-inflammatory drugs, ibuprofen, from sandy soils. About 54–66% removal was obtained when nZVI NPs were utilized in aqueous solution, whereas about 95% degradation was obtained when nZVI was utilized as catalyst in the Fenton process to remove ibuprofen. Fazlzadeh et al. (2017) prepared zero-valent iron nanoparticles (ZVI NPs) of irregular morphology with an average particle size of 100 nm from three plant extracts, inclusive of *Rosa damascene* (RD), *Thymus vulgaris* (TV), and *Urtica dioica* (UD) for the removal of hexavalent chromium in aqueous phase.

Patra and Baek (2017) reported the formation of magnetic iron oxide nanoparticles from the extracts obtained from (*Zea mays* L.) and outer leaves of Chinese cabbage (*Brassica rapa L. subsp. pekinensis*) via a photo-catalyzed process. The resultant Fe NPs exhibited relatively high antibacterial, anticandidal, and antioxidant activity. Prasad et al. (2017a, 2017b, 2017c, 2017d) conducted several studies to produce functionalized magnetic iron oxide nanoparticles from leaf extracts of four plant species, namely pomegranate, *Murrayakoenigii*, *Moringa oleifera*, and *Pisum sativum* to remove lead and organic dyes, like congo red, malachite green and methyl orange from wastewater. The particle size of the NPs varied according to the plant extract. The authors reported the formation of spherical 12 nm reduced graphene oxide – magnetic iron oxide nanocomposites from the *Murraya koenigii* leaf extract. In contrast, rod-shaped Fe_3O_4 NPs with a uniform size distribution of 45–60 nm and a high specific surface area of 14.6 m^2/g were synthesized using the extract from pomegranate leaf.

Khalil et al. (2017) utilized leaves of *Sageretia thea* for synthesizing tetragonal crystalline-shaped iron oxide nanoparticles. The average nanoparticle size was found to be ~29 nm. The bioactive components were extracted from pulverized plant leaves with deionized water which further acted as reducing agent. The synthesized nanoparticles were utilized to study the antibacterial effect against five human pathogenic bacterial strains, and out of which *Pseudomonas aeruginosa* was found to be most susceptible.

Dried green tea was reported by Borja et al. (2015) for synthesizing nano zero-valent iron. Green tea leaves composed of abundant polyphenols, and bioactive compounds which were extracted through microwave assistance. The obtained nanoparticles were spherical in nature and sizes ranged from 8 to 23 nm. Ethanol was utilized as the extractant solvent for extracting the bioactive components. Maximum nanoparticle yield was investigated to depend on factors like, precursor:plant extract ratio, plant material:solvent ratio, and plant extract delivery rate into precursor solution. Further, Wei et al. (2017) synthesized zero-valent iron NPs utilizing extract of *Eichhornia crassipes* (a water hyacinth). Figure 4.2 shows that the nanoparticles were spherical-shaped, amorphous in nature with average particle size of 20–80 nm. The prepared nanoparticles were utilized for Cr (VI) removal study. Through batch experiments, 89.9% of Cr (VI) removal was obtained by the Ec-Fe-NPs which were found to be much higher than the removal obtained by extracts (20.4%) and Fe_3O_4 nanoparticles (47.3%) alone.

Prabhakar and Samadder (2017) similarly reported the utilization of *Eichhornia crassipes, Lantana camara,* and *Mimosa pudica* leaves for the synthesis of iron nanoparticles. Extract of *Eichhornia crassipes* leaves synthesized spherical iron nanoparticles with average particle size of 20–60 nm. Irregular and aggregated quasispherical morphology of nanoparticles were obtained from *L. Camara* and

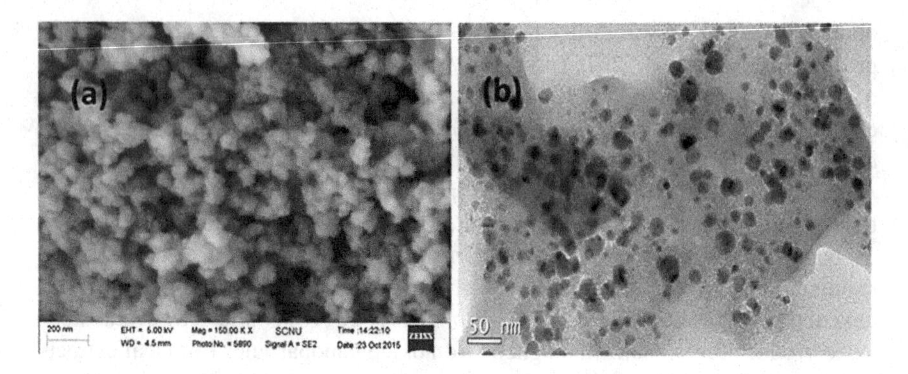

FIGURE 4.2 (a) SEM and (b) TEM images of the synthesized zero-valent iron NPs [Reproduced with permission from Wei et al. (2017) © Elsevier].

M. pudica leaf extract. The synthesized nanoparticles were utilized for nitrate removal study and compared with chemical synthesized iron NPs. *Eichhornia crassipes, Lantana camara* and *Mimosa pudica* mediated iron nanoparticles showed 74.52%, 71.12% and 65.23% of nitrate removal respectively, whereas 85.27% removal was obtained for chemically synthesized iron NPs.

Coriandrum sativum leaves were utilized by Sathya et al. (2017) for synthesizing iron oxide nanoparticles. The synthesized nanoparticles were spherical in nature with average size ranging from 20 to 90 nm. Continuous stirring and ultrasonication techniques were utilized for synthesis. Shorter synthesis time was obtained for ultrasonication method, which wa further utilized for antimicrobial activity test. *Micrococcus luteus* and *Staphylococcus aureus* shows higher antimicrobial activity with mean zone diameter of 22 mm and 18 mm respectively.

Singh et al. (2017) synthesized uniform shaped 10±3nm iron oxide nanoparticles from green tea leaves extract. The synthesized iron oxide nanoparticles were superparamagnetic in nature with magnetic saturation 61 emu/g. Further, the synthesized nanoparticle was applied for methylene blue dye removal. A maximum adsorption capacity of 7.25 mg/g was obtained with reaction kinetic following pseudo-second-order along with the Langmuir isotherm model.

Neem leaf extract was utilized by Madhubala and Kalaivani (2017) for preparing iron oxide nanoparticles supported on zinc oxide. The synthesized iron oxide nanoparticles were found to be brick-like structures with an average particle size of 38 nm. Subsequently, the synthesized iron oxide nanoparticles were mixed with zinc oxide through ultrasonication, and further autoclaved for 18 h at 200°C. The nanocomposite prepared was found to be ellipsoid-shaped with average size ranging from 45 to 61 nm. Vero cells were used for studying the toxicity of the Fe_3O_4@ZnO core-shell nanoparticles using MTT assay. Results show that with higher concentration of nanoparticles the cell viability decreases.

Harshiny et al. (2015) synthesized both chemical synthesized iron oxide nanoparticles along with *Amaranthus dubius* leaf extract mediated nanoparticles. Various photochemicals present in the leaf extract such as flavonoids, phenols, and lysine generated higher zeta potential about –44 mV to –66 mV with better stability than chemical synthesized iron nanoparticles. Particle size analysis showed green synthesized NPs had particle size ranging between 43 and 220 nm, whereas chemical synthesized nanoparticles varied between 70 and 700 nm. Morphological analysis showed green synthesized nanoparticles had spherical structures with less aggregation and particle size ranging from 60 to 300 nm. Furthermore, both synthesized nanoparticles were utilized for degradation of methyl orange (MO). The degradation efficiency was found to be higher for chemical synthesized nanoparticles (90%), whereas green synthesized nanoparticles had an efficiency of 81%, under experimental conditions of 20 ppm MO, and 20 mg of NPs.

Murraya koenigii (curry leaves) leaf extract was utilized by Mohanraj et al. (2014) for synthesizing iron oxide NPs. The synthesized nanoparticles were spherical in nature with irregular aggregates having an average size of ~59 nm. Lower particle size distribution was observed due to capping agents from the extract which consist of alkaloids, flavonoids, carbazole, and polyphenols. Moreover, the synthesized Fe NPs were utilized for hydrogen production from glucose by *Clostridium acetobutylicum* NCIM 2337. During controlled experiments 1.74 ± 0.08 mol H_2/mol glucose of hydrogen yield was observed, whereas 175 mg/L Fe NPs mediated process achieved hydrogen yield of 2.33 ± 0.09 mol H_2/mol glucose. Results conveyed that increase in hydrogen content from 34 ± 0.8 to 52 ± 0.8 % and hydrogen production rate from 23 to 25.3 mL/h, were obtained.

Senthil and Ramesh (2012) reported the antibacterial effect of *Tridox procumbens* leaf extract mediated iron nanoparticles. Due to the higher carbohydrate content of the leaf it acted as an efficient reducing agent. Morphological analysis showed irregular spherical shaped iron oxide NPs with rough surface. The crystalline structure of Fe_3O_4 was confirmed by XRD analysis which confirmed the crystalline size ranging from 80 to 100 nm. *Pseudomonas aeruginosa* was utilized for studying the antibacterial activity of green synthesized iron oxide NPs using agar well diffusion method, and results showed its efficient antibacterial activity.

Another synthesis method was employed by Kiruba Daniel et al. (2013), where *Dodonaea viscosa* leaf extract was utilized to synthesize Fe NPs. During the experiment 5 mL of leaf extracted was added to 10 mL of 10 mM $FeCl_3$ solution. Immediate color change indicated formation of nZVI nanoparticles. The synthesized nanoparticles were found to be spherical in nature with average diameter within 50–60 nm. The rich content of polyphenols such as tannins, santin, saponins, and Pendleton in *Dodonaea viscosa* leaf extract were found useful in synthesizing nanoparticles with stabilized capping.

Naseem and Farrukh (2015) reported synthesis of Fe NPs using *Lawsonia inermis* (henna) and *Gardenia jasminoides* leaf extract by simple conventional heating method. The average diameter of Fe NPs obtained using *Lawsonia inermis* leaf extract was 21 nm, whereas an average diameter of 32 nm was obtained for *Gardenia* leaf extract mediated Fe NPs. Through experiment it was found that lawsone (2-hydroxy-1,4-naphthoquinone) was the main component within henna, which basically consists of units such as p-benzoquinone, benzene, and phenols.

Hence such bioactive components synthesized better stabilized coated Fe NPs. Henna synthesized Fe NPs were found to be agglomerated and had distorted hexagonal-like shape. Whereas shattered rod-like Fe NPs were observed in agglomerated form using *Gardenia*, leaf extract. The synthesized Fe NPs, due to better bioactive coatings, were found to show better antibacterial activity against human pathogens such as *Escherichia coli* and *Staphylococcus aureus*.

Moreover, nano zero-valent iron (nZVI) NPs were synthesized by Kumar et al. (2015) utilizing *Emblica officinalis* leaf extract. The high content of polyphenol and ascorbic acid reduced the metal ions and further formed stabilizing agents over nZVI NPs. Morphological analysis revealed spherical morphology of the nZVI NPs with an average diameter of 22.6 nm. Zeta potential value of −26.7 mV indicated moderate stabilization of nZVI NPs. Further, the nanoparticles were utilized for Pb(II) removal study. A maximum removal of 99% was obtained for 10ppm Pb(II) solution, at the end of 24 h with adsorbent dosage of 2 mL/L.

In a similar study, Wang et al. (2014b) utilized *Eucalyptus* leaves and through a very simple, and efficient technique synthesized iron oxide NPs. Experimental studies involved mixing of the leaf extract and 0.10 M $FeSO_4$ in a volume ratio of 2:1. The solutional color change from yellow to black signified formation of iron oxide NPs. The NPs were found to be spherical in nature with average diameter in the range of 20–80 nm. Further the synthesized iron NPs were utilized towards remediation of swine wastewater, and was found to remove 71.7% of total N and 84.5% of COD.

4.2 FRUIT EXTRACT

The reduction potential of various plant extracts plays an important role for the selection to be utilized for nanoparticle synthesis. Researchers need to keep this in mind while exploring various plant materials and need to choose extracts with most bioactive components for efficient nanoparticle synthesis having better capping and stabilization and further it will lead to their appropriate application in various fields. Mohan Kumar et al. (2013) utilized *Terminalia chebula* fruit extract, a plant rich in polyphenolic content, and studied its redox potential through cyclic voltammetry. A sufficient reduction potential of 0.63 V was obtained with respect to *Camellia sinensis* (green tea), with redox potential of 0.33 V. Morphology analysis showed that chain-like iron NPs were formed in agglomerate form with an average particle diameter of 80 nm.

Saikia et al. (2017) reported the utilization of fruit extract and rice paddy husk for the synthesis of silica-supported iron oxide nanocomposite. In the first step, *Zanthoxylum rhetsa* fruit extract was employed for synthesizing iron oxide nanoparticles through iron (III) solution as precursor. The second step comprised of utilizing rice straw for preparing silica by boiling the raw material with sodium hydroxide solution and sulfuric acid. Finally iron oxide-silica composite was prepared by mixing the prepared raw materials and refluxing with methanol. Figure 4.3 shows the average spherical particle diameter of 5–21 nm which was obtained for the prepared NPs. Further the prepared Fe_2O_3@silica nanocomposite

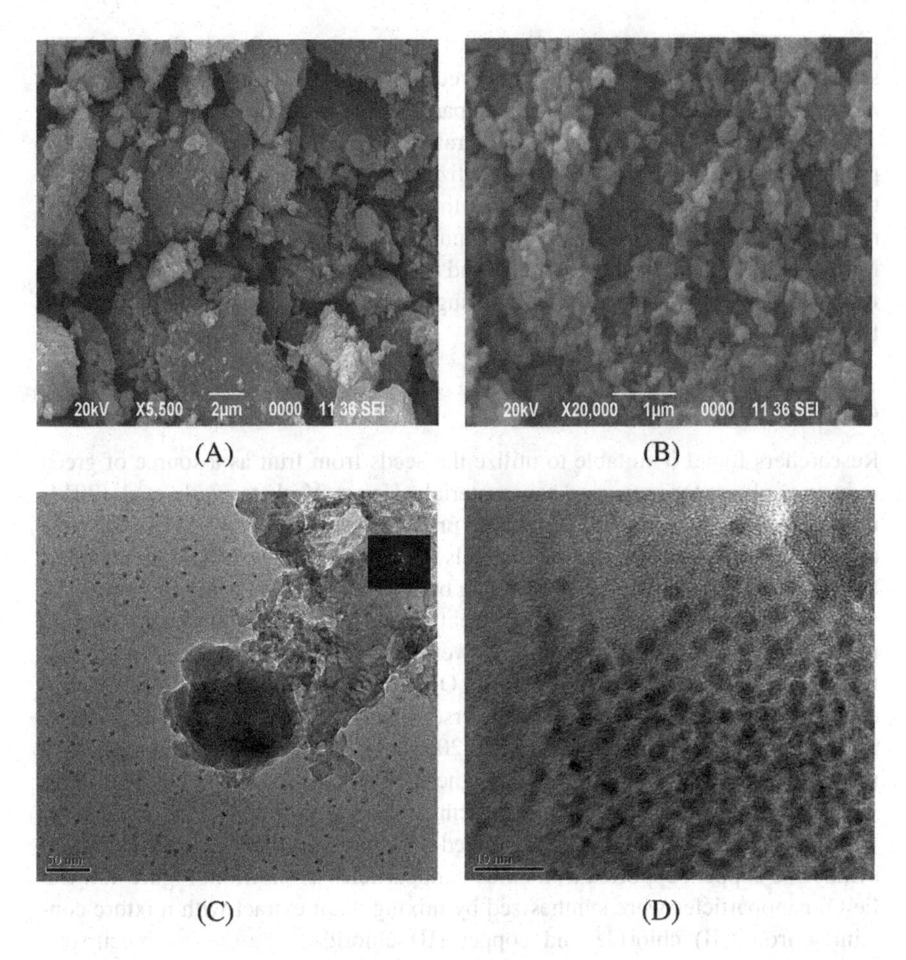

FIGURE 4.3 (a, b) SEM (C-D) TEM images of $Fe_2O_3@SiO_2$ nanocomposite [Reproduced with permission from Saikia et al. (2017) © Elsevier].

was utilized for ipso-hydroxylation of phenylboronic acid to phenols. A maximum yield of 98% was obtained at the end of 120 min in the presence of H_2O_2 as catalyst.

Blueberries were utilized by Manquián-Cerda et al. (2017) as a source for green synthesis of nZVI NPs. An agglomerated form was obtained with an average particle diameter of 52.4 nm. Ferric reducing antioxidant power along with total phenolic content of the prepared fruit extract showed the capability of forming better stabilized iron NPs. The surface area of the synthesized nanoparticle was found to be 70.7 m^2/g. The synthesized NPs was utilized for removing As(V) and was found to have sorption capacity of 52.23 ± 6.06 mg/g.

Similarly, watermelon rinds were utilized by Prasad et al. (2015) for preparing Fe_3O_4 magnetic nanoparticles. Watermelon rinds were non-toxic and biodegradable in nature, and were found to be rich in polyphenol, acid derivatives, and protein. The presence of such functional groups was confirmed through FTIR spectra. Morphological analysis conveyed that the particles were crystalline in nature, and spherical in shape. Average particle diameter was found to be 20 nm. The magnetic nanoparticle showed saturation magnetization of 14.2 emu/g. The prepared magnetic Fe_3O_4 NPs were utilized for the synthesis of 2-oxo-1,2,3,4-tetrahydropyrimidine compounds due to its catalytic property. Experimental results showed at a catalyst loading of 5 mmol highest yield of 94% was obtained. Moreover, the magnetic nanoparticles had the advantage of being easily separated from the process product formed by using a strong magnet and can be recycled further.

4.3 SEED EXTRACT

Researchers found it suitable to utilize the seeds from fruit as a source of green reducing solvent for preparing nanomaterials. Hence, Venkateswarlu et al. (2014) utilized *Syzygium cumini* seed extract to prepared magnetic Fe_3O_4-NPs. The huge content of carbohydrates and polyphenols helped in synthesizing Fe_3O_4 nanoparticles by simple reduction reaction acting both as reducing and capping agent. The synthesized nanomaterial showed excellent ferromagnetic behavior with saturation magnetization value of 13.6 emu/g. However, Raman spectroscopy characterization confirmed the formation of pure Fe_3O_4 without any impurity. Morphological characteristics conveyed crystalline inverse spinel cubical structure with an average particle diameter ranging from 9 to 20 nm. An average surface area of 3.517 m^2/g was obtained. Such prepared magnetic nanoparticle was believed to have wide application in the field of biomedicine and separation aspects.

Moreover, Sajadi et al. (2016) utilized *Silybum marianum L.* seed extract to synthesize copper supported iron oxide nanoparticles as shown in Figure 4.4. Cu/Fe_3O_4 nanoparticles were synthesized by mixing plant extract with mixture containing iron (III) chloride and copper (II) chloride solutions as precursors. Morphological analysis confirmed the synthesis of crystalline spherical nanoparticles with an average particle diameter ranging from 8.5 to 60 nm. Further the

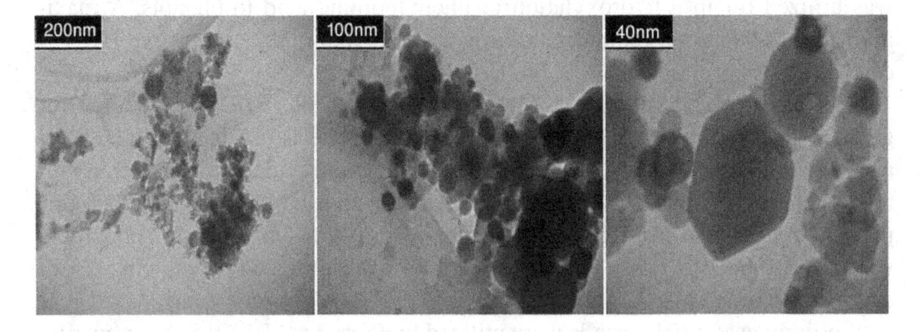

FIGURE 4.4 TEM images of the green synthesized Cu/Fe_3O_4 NPs obtained utilizing *Silybum marianum L.* seed extract [Reproduced with permission from Sajadi et al. (2016) © Elsevier].

prepared nanomaterial was utilized for catalytic reduction of nitroarenes in ethyl alcohol solution in presence of sodium borohydride as catalyst. Experimental study reveals that a maximum of 95% yield was obtained in the first cycle, whereas the yield reduced to 92% at the end of fifth recycle. This suggests the practical reusability of the synthesized nanomaterial in catalytic application purpose.

4.4 STOLON AND ROOT

Recent study suggests that few research works are available where the part of plants such as stolon (potato), root and gum have not been utilized for synthesizing Fe_3O_4-NPs. Buazar et al. utilized starch rich potato extract for synthesizing Fe_3O_4-NPs. Starch content extract plays an effective role in reducing along with stabilizing the Fe_3O_4-NPs with capping agents mainly consisting of carbohydrate groups. The oxidation of starch extract was started with the addition of NaOH, which further produced electrons for producing Fe^0 nanoparticles. Crystalline structure of magnetite NPs were formed with face centered cubic shape having an average particle diameter of 40 ± 2.2 nm, as shown in Figure 4.5. Furthermore, the synthesized magnetite NPs was utilized for catalytic degradation of organic methylene blue contaminant at room temperature.

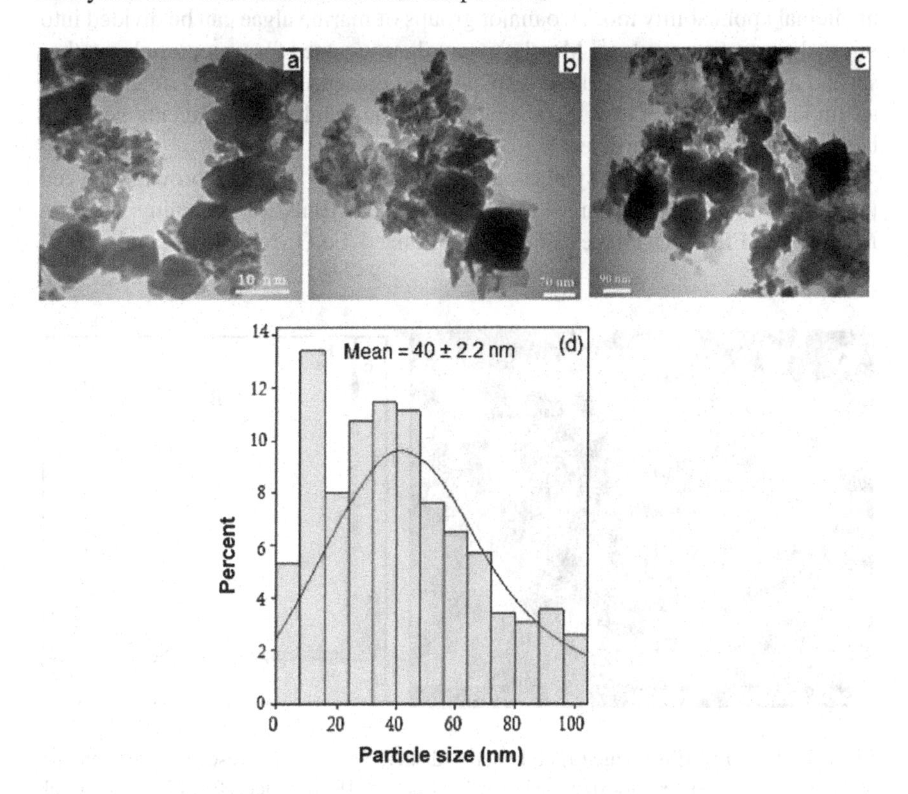

FIGURE 4.5 TEM image (a–c) and particle size distributions (d) of biosynthesized Fe_3O_4 NPs [Reproduced with permission from Buazar et al. (2016) © Wiley & Sons].

Niraimathee et al. (2016) by using *Mimosa pudica* root extract reported the production of Fe_3O_4-NPs. The synthesized iron oxide NPs showed a sharp absorbance peak of 294 nm through UV analysis. The superparamagnetic behavior of the Fe_3O_4-NPs were indicated by the saturation magnetization value which was found to be ~55.40 emu/g, considerably higher than reported values. Morphological analysis showed that the synthesized NPs were spherical in shape with an average particle diameter of 67 nm. Single crystal domain of the magnetic NPs were observed which signifies single orientation of magnetic moment. The magnetization value was found to decrease to zero from the plateau value on removal of the magnetic field.

4.5 MARINE PLANTS

Marine plants, usually known as seaweed or algae, were also considered to be useful as a source for synthesizing Fe_3O_4-NPs. Seaweed is found in abundance in south-east Asia and is considered vital for marine organisms since it provides food and habitat. Due to the presence of abundant lipids, minerals, and some vitamins, it is considered as a source of food. Moreover, due to the presence of various phytochemicals such as proteins, polyphenols, and polysaccharides it has medicinal applicability too. Two major groups of marine algae can be divided into microalgae and macroalgae. Mostly, macroalgae are used for nanoparticle synthesis and are basically plant-like organisms. The presence of phytochemicals plays an important role, as these metal-reducing agents and capping agents generate better stabilized nanoparticles.

Mahdavi et al. (2013) synthesized Fe_3O_4-NPs by utilizing brown seaweed (*Sargassum muticum*). Figure 4.6 shows the Field Emission Scanning Electron Microsope (FESEM) images of the synthesized Fe_3O_4-NPs. The synthesis of

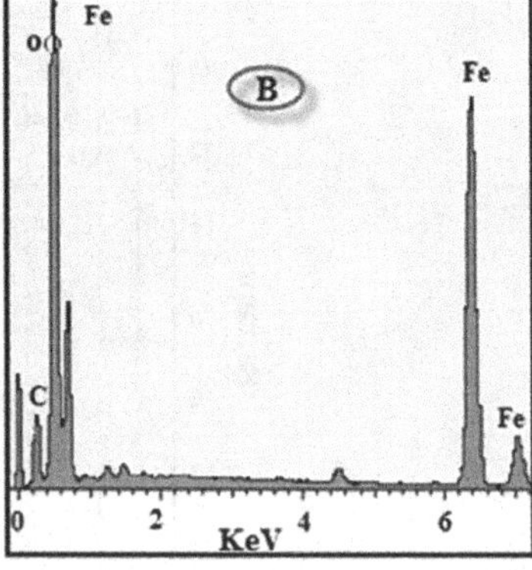

FIGURE 4.6 FESEM image (a) and energy-dispersive X-ray fluorescence spectrometry spectra of Fe_3O_4-NPs synthesized using BS extract (b) [Reproduced from Mahdavi et al. (2013) © MDPI].

Fe_3O_4-NPs was achieved by simple mixing of $FeCl_3$ solution and the extract of brown seaweed. After mixing a few minutes were required to initiate the reaction and after that, Fe_3O_4-NPs were immediately produced. The reduction of Fe^{3+} ions and surface coating for stabilization was carried out by the presence of sulphate, hydroxyl and aldehyde group in seaweed. The participation of –OH group in the reaction was confirmed by the reduction of the solution pH. Due to the oxidation of aldehyde groups to carboxylic acid molecules by sulphate group, hence it was suggested that the reduction of metal ions was performed by sulphate ions. The synthesized Fe_3O_4-NPs was found to be cubical in shape with an average diameter of 18 ± 4 nm, with a saturation magnetization value of 22.1 emu/g, and negligible coercivity signifying the superparamagnetic nature of Fe_3O_4-NPs.

REFERENCES

Al-Ruqeishi, M.S., Mohiuddin, T., Al-Saadi, L.K. 2016. Green synthesis of iron oxide nanorods from deciduous Omani mango tree leaves for heavy oil viscosity treatment. *Arab. J. Chem.* 12 (8), 4084–4090.

Afsheen, S., Bilal, M., Iqbal, T., Liaqat, A., Abrar, M. 2018. Green synthesis and characterization of novel iron particles by using different extracts. *J. Alloys Compd.* 732, 935–944.

Ahmed, M.J.K., Ahmaruzzaman, M., Bordoloi, M.H. 2015. Novel Averrhoa carambola extract stabilized magnetite nanoparticles: A green synthesis route for the removal of chlorazol black E from wastewater. *RSC Adv.* 5, 74645–74655.

Atarod, M., Nasrollahzadeh, M., Mohammad Sajadi, S. 2015. Green synthesis of Pd/RGO/Fe_3O_4 nanocomposite using *Withania coagulans* leaf extract and its application as magnetically separable and reusable catalyst for the reduction of 4-nitrophenol. *J. Colloid Interface Sci.* 465, 249–258.

Awwad, A.M.; Salem, N.M., 2012. A green and facile approach for synthesis of magnetite nanoparticles. *Nanosci. Nanotechnol.* 2, 208–213.

Bahadur, A., Saeed, A., Shoaib, M., Iqbal, S., Bashir, M.I., Waqas, M., Hussain, M.N., Abbas, N., 2017. Eco-friendly synthesis of magnetite (Fe_3O_4) nanoparticles with tunable size: Dielectric, magnetic, thermal and optical studies. *Mat. Chem. Phys.* 198, 229–235.

Basavegowda, N., Magar, K.B.S., Mishra, K., Lee, Y.R., 2014a. Green fabrication of ferromagnetic Fe_3O_4 nanoparticles and their novel catalytic applications for the synthesis of biologically interesting benzoxazinone and benzthioxazinone derivatives. *New J. Chem.* 38, 5415–5420.

Basavegowda, N., Mishra, K., Lee, Y.R., 2014b. Sonochemically synthesized ferromagnetic Fe_3O_4 nanoparticles as a recyclable catalyst for the preparation of pyrrolo[3, 4-c] quinoline-1,3-dione derivatives. *RSC Adv.* 4, 61660–61666.

Borja, J.Q., Ngo, M.A.S., Saranglao, C.C., Tiongco, R.P.M., Roque, E.C., Dugos, N.P. 2015. Synthesis of green zero-valent iron using polyphenols from dried green tea extract. *J. Eng. Sci. Technol.* 10 (Spec.issue7), 22–31.

Buazar, F., Baghlani-Nejazd, M.H., Badri, M., Kashisaz, M., Khaledi-Nasab, A., Kroushawi, F. 2016. Facile one-pot phytosynthesis of magnetic nanoparticles using potato extract and their catalytic activity. *Starch-Starke* 68, 796–804.

Cai, Y., Shen, Y., Xie, A., Li, S., Wang, X., 2010. Green synthesis of soya bean sprouts-mediated superparamagnetic Fe_3O_4 nanoparticles. *J. Mag. Mag. Mat.* 322 (19), 2938–2943.

Dhillon, G.S., Brar, S.K., Kaur, S., Verma, M. 2012. Green approach for nanoparticle biosynthesis by fungi: Current trends and applications. *Crit. Rev. Biotechnol.* 32 (1), 49–73.

El-Kassas Hala, Y., Aly-Eldeen Mohamed, A., Gharib Samiha, M. 2016. Green synthesis of iron oxide (Fe_3O_4) nanoparticles using two selected brown seaweeds: Characterization and application for lead bioremediation. *Acta Oceanol. Sin.* 35, 89–98.

Fazlzadeh, M., Rahmani, K., Zarei, A., Abdoallahzadeh, H., Nasiri, F., Khosravi, R. 2017. A novel green synthesis of zero valent iron nanoparticles (NZVI) using three plant extracts and their efficient application for removal of Cr(VI) from aqueous solutions. *Adv. Powder Technol.* 28 (1), 122–130.

Gan, L., Lu, Z., Cao, D., Chen, Z. 2018. Effects of cetyltrimethylammonium bromide on the morphology of green synthesized Fe_3O_4 nanoparticles used to remove phosphate. *Mater. Sci. Eng. C.* 82, 41–45.

Groiss, S., Selvaraj, R., Varadavenkatesan, T., Vinayagam, R. 2017. Structural characterization, antibacterial and catalytic effect of iron oxide nanoparticles synthesized using the leaf extract of *Cynometra ramiflora. J. Mol. Struct.* 1128, 572–578.

Harshiny, M., Iswarya, C.N., Matheswaran, M. 2015. Biogenic synthesis of iron nanoparticles using *Amaranthus dubius* leaf extract as a reducing agent. *Pow. Technol.* 286, 744–749.

Herrera-Becerra, R., Zorrilla, C., Ascencio, J.A. 2007. Production of iron oxide nanoparticles by a biosynthesis method: An environmentally friendly route. *J. Phys. Chem.* 111, 16147–16153.

Huang, L., Luo, F., Chen, Z., Megharaj, M., Naidu, R. 2015. Green synthesized conditions impacting on the reactivity of Fe NPs for the degradation of malachite green. *Spectrochim. Acta A* 137, 154–159.

Iravani, S. 2011. Green synthesis of metal nanoparticles using plants. *Green Chem.* 13 (10), 2638–2650.

Jagathesan, G., Rajiv, P. 2018. Biosynthesis and characterization of iron oxide nanoparticles using *Eichhornia crassipes* leaf extract and assessing their antibacterial activity. *Biocatal. Agric. Biotechnol.* 13, 90–94.

Jha, P.K. 2017. Green synthesis of magnetic Fe_3O_4 nanoparticles using *Couroupita guianensis* Aubl. fruit extract for their antibacterial and cytotoxicity activities. *Artif. Cells Nanomed. Biotechnol.* 2017, 1–10.

Kalaiarasi, R., Jayallakshmi, N., Venkatachalam, P. 2010. Phytosynthesis of nanoparticles and its applications. *Plant Cell Biotechnol Mol Biol.* 11 (1/4), 1–16.

Kanagasubbulakshmi, S., Kadirvelu, K. 2017. Green synthesis of Iron oxide nanoparticles using *Lagenaria siceraria* and evaluation of its Antimicrobial activity. *Def. Life Sci. J.* 2, 422–427.

Khaghani, S., Ghanbari, S.K. 2017. Green synthesis of Iron oxide-Palladium nanocomposites by pepper extract and its application in removing of colored pollutants from water. *J. Nano. Chem.* 7 (3), 175–182.

Khalil, A.T., Ovais, M., Ullah, I., Ali, M., Khan Shinwari, Z., Maaza, M. 2017. Biosynthesis of iron oxide (Fe_2O_3) nanoparticles via aqueous extracts of *Sageretia thea* (Osbeck.) and their pharmacognostic properties. *Green Chem. Lett. Rev.* 10 (4), 186–201.

Kiruba Daniel, S.C.G., Vinothini, G., Subramanian, N., Nehru, K., Sivakumar, M.M. 2013. Biosynthesis of Cu, ZVI, and Ag nanoparticles using *Dodonaea viscosa* extract for antibacterial activity against human pathogens. *J. Nano. Res.* 15, 1–10.

Kumar, R., Singh, N., Pandey, N. 2015. Potential of green synthesized zero-valent iron nanoparticles for remediation of lead-contaminated water. *Sci. Technol.* 12, 3943–3950.

Latha, N., Gowri, M. 2014. Biosynthesis and characterisation of Fe_3O_4 nanoparticles using Caricaya papaya leaves extract. *Int. J. Sci. Res.* 3, 1551–1556.

Machado, S., Pacheco, J.G., Nouws, H.P.A., Albergaria, J.T., Delerue-Matos, C. 2017. Green zero-valent iron nanoparticles for the degradation of amoxicillin. *Int. J. Environ. Sci. Technol.* 14 (5), 1109–1118.

Machado, S., Pinto, S.L., Grosso, J.P., Nouws, H.P.A., Albergaria, J.T., Delerue-Matos, C. 2013. Green production of zero-valent iron nanoparticles using tree leaf extracts. *Sci. Tot. Environ.* 445, 1–8.

Madhubala, V., Kalaivani, T. 2017. Phyto and hydrothermal synthesis of Fe_3O_4@ZnO core-shell nanoparticles using *Azadirachta indica* and its cytotoxicity studies. *Appl. Surf. Sci.* 449, 584–590.

Mahdavi, M., Namvar, F., Ahmad, M.B., Mohamad, R. 2013. Green biosynthesis and characterization of magnetic iron oxide (Fe_3O_4) nanoparticles using seaweed (*Sargassum muticum*) aqueous extract. *Molecules* 18 (5), 5954–5964.

Manquian-Cerda, K., Cruces, E., Angelica Rubio, M., Reyes, C., Arancibia-Miranda, N. 2017. Preparation of nanoscale iron (oxide, oxyhydroxides and zero-valent) particles derived from blueberries: Reactivity, characterization and removal mechanism of arsenate. *Ecotoxicol. Environ. Saf.* 145, 69–77.

Mohan Kumar, K., Mandal, B.K., Siva Kumar, K., Sreedhara Reddy, P., Sreedhar, B. 2013. Biobased green method to synthesise palladium and iron nanoparticles using *Terminalia chebula* aqueous extract. *Spectro. Acta A* 102, 128–133.

Mohanraj, S., Kodhaiyolii, S., Rengasamy, M., Pugalenthi, V. 2014. Green synthesized iron oxide nanoparticles effect on fermentative hydrogen production by Clostridium acetobutylicum. *App. Biochem. Biotechnol.* 173, 318–331.

Mondal, P., Anweshan, A., Purkait, M.K. 2020. Green synthesis and environmental application of iron-based nanomaterials and nanocomposite: A review. *Chemosphere* 259, 127509.

Mukunthan, K.S., Balaji, S. 2012. Silver nanoparticles shoot up from the root of *Daucus carrota* (L.). *Inter. J. Green Nanotech.* 4 (1), 54–61.

Narayanan, S., Sathy, B.N., Mony, U., Koyakutty, M., Nair, S.V., Menon, D. 2011. Biocompatible magnetite/gold nanohybrid contrast agents via green chemistry for MRI and CT bioimaging. *ACS Appl. Mat. Int.* 4, 251–260.

Naseem, T., Farrukh, M.A. 2015. Antibacterial activity of green synthesis of iron nanoparticles using *Lawsonia inermis* and *Gardenia jasminoides* leaves extract. *J. Chem.* 2015, 1–7.

Ngernpimai, S., Thomas, C., Maensiri, S., Siri, S. 2012. Stability and cytotoxicity of well-dispersed magnetite nanoparticles prepared by hydrothermal method. *Adv. Mat. Res.* 506, 122–125.

Niraimathee, V., Subha, V., Ravindran, R.E., Renganathan, S. 2016. Green synthesis of iron oxide nanoparticles from *Mimosa pudica* root extract. *Int. J. Environ. Sustain. Dev.* 15, 227–240.

Patra, J.K., Baek, K.H. 2017. Green biosynthesis of magnetic iron oxide (Fe_3O_4) nanoparticles using the aqueous extracts of food processing wastes under photo-catalyzed condition and investigation of their antimicrobial and antioxidant activity. *J. Photochem. Photobio. B: Biol.* 173, 291–300.

Phumying, S., Labuayai, S., Thomas, C., Amornkitbamrung, V., Swatsitang, E., Maensiri, S. 2012. Aloe vera plant-extracted solution hydrothermal synthesis and magnetic properties of magnetite (Fe_3O_4) nanoparticles. *Appl. Phys. A* 111, 1187–1193.

Prabhakar, R., Samadder, S.R. 2017. Jyotsana Aquatic and terrestrial weed mediated synthesis of iron nanoparticles for possible application in wastewater remediation. *J. Clean. Prod.* 168, 1201–1210.

Prasad, A.S. 2016. Iron oxide nanoparticles synthesized by controlled bio-precipitation using leaf extract of Garlic Vine (*Mansoa alliacea*). *Mat. Sci. Semi. Proc.* 53, 79–83.

Prasad, C., Gangadhara, S., Venkateswarlu, P. 2015. Bioinspired green synthesis of Fe_3O_4 magnetic nanoparticles using watermelon rinds and their catalytic activity. *App. Nanosci.* 5, 847–855.

Prasad, C., Karlapudi, S., Venkateswarlu, P., Bahadur, I., Kumar, S. 2017a. Green arbitrated synthesis of Fe_3O_4 magnetic nanoparticles with nanorod structure from pomegranate leaves and Congo red dye degradation studies for water treatment. *J. Mol. Liq.* 240, 322–328.

Prasad, C., Krishna Murthy, P., Hari Krishna, R.H., Sreenivasa Rao, R., Suneetha, V., Venkateswarlu, P. 2017b. Bio-inspired green synthesis of RGO/Fe_3O_4 magnetic nanoparticles using Murrayakoenigii leaves extract and its application for removal of Pb (II) from aqueous solution. *J. Environ. Chem. Eng.* 5 (5), 4374–4380.

Prasad, C., Sreenivasulu, K., Gangadhara, S., Venkateswarlu, P. 2017c. Bio inspired green synthesis of Ni/Fe_3O_4 magnetic nanoparticles using *Moringa oleifera* leaves extract: A magnetically recoverable catalyst for organic dye degradation in aqueous solution. *J. Alloys. Compd.* 700, 252–258.

Prasad, C., Yuvaraja, G., Venkateswarlu, P. 2017d. Biogenic synthesis of Fe_3O_4 magnetic nanoparticles using *Pisum sativum* peels extract and its effect on magnetic and Methyl orange dye degradation studies. *J. Magn. Magn. Mat.* 424, 376–381.

Ramasahayam, S.K., Gunawan, G., Finlay, C., Viswanathan, T. 2012. Renewable resource-based magnetic nanocomposites for removal and recovery of phosphorous from contaminated water. *Water, Air, Soil Poll.* 223, 4853–4863.

Rao, A., Bankar, A., Kumar, A.R., Gosavi, S., Zinjarde, S. 2013. Removal of hexavalent chromium ions by Yarrowia lipolytica cells modified with phyto-inspired Fe^0/Fe_3O_4 nanoparticles. *J. Contam. Hydrol.* 146, 63–73.

Rajendran, S.P., Sengodan, K. 2017. Synthesis and characterization of zinc oxide and iron oxide nanoparticles using *Sesbania grandiflora* leaf extract as reducing agent. *J. Nanosci.* 2017, 1–7.

Rajiv, P., Bavadharani, B., Kumar, M.N., Vanathi, P. 2017. Synthesis and characterization of biogenic iron oxide nanoparticles using green chemistry approach and evaluating their biological activities. *Biocatal. Agric. Biotechnol.* 12, 45–49.

Saikia, I., Hazarika, M., Hussian, N., Das, M.R., Tamuly, C. 2017. Biogenic synthesis of Fe_2O_3@SiO_2 nanoparticles for ipso-hydroxylation of boronic acid in water. *Tetrahedron Lett.* 58 (45), 4255–4259.

Sajadi, S.M., Nasrollahzadeh, M., Maham, M. 2016. Aqueous extract from seeds of *Silybum marianum* L. as a green material for preparation of the Cu/Fe_3O_4 nanoparticles: A magnetically recoverable and reusable catalyst for the reduction of nitroarenes. *J. Colloid Interface Sci.* 469, 93–98.

Sangami, S., Manu, B. 2017. Synthesis of green iron nanoparticles using laterite and their application as a Fenton-like catalyst for the degradation of herbicide Ametryn in water. *Environ. Technol. Innov.* 8, 150–163.

Sathya, K., Saravanathamizhan, R., Baskar, G. 2017. Ultrasound assisted phytosynthesis of iron oxide nanoparticle. *Ultrason. Sonochem.* 39, 446–451.

Senthil, M., Ramesh, C. 2012. Biogenic synthesis of Fe_3O_4 nanoparticles using *Tridax procumbens* leaf extract and its antibacterial activity on Pseudomonas aeruginosa. *Dig. J. Nanomat. Biostruc.* 7, 1655–1660.

Silveira, C., Shimabuku, Q.L., Fernandes Silva, M., Bergamasco, R. 2017. Iron-oxide nanoparticles by green synthesis method using *Moringa oleifera* leaf extract for fluoride removal. *Environ. Technol.* 3330, 1–40.

Singh, K.K., Senapati, K.K., Sarma, K.C. 2017. Synthesis of superparamagnetic Fe_3O_4 nanoparticles coated with green tea polyphenols and their use for removal of dye pollutant from aqueous solution. *J. Environ. Chem. Eng.* 5 (3), 2214–2221.

Sirdeshpande, K.D., Sridhar, A., Cholkar, K.M., Selvaraj, R. 2018. Structural characterization of mesoporous magnetite nanoparticles synthesized using the leaf extract of *Calliandra haematocephala* and their photocatalytic degradation of malachite green dye. *Appl. Nanosci.* 2018, 1–9.

Venkateswarlu, S., Kumar, B.N., Prasad, C., Venkateswarlu, P., Jyothi, N. 2014. Bio-inspired green synthesis of Fe_3O_4 spherical magnetic nanoparticles using *Syzygium cumini* seed extract. *Physica. B. Condens. Mat.* 449, 67–71.

Wang, T., Jin, X., Chen, Z., Megharaj, M., Naidu, R. 2014b. Green synthesis of Fe nanoparticles using eucalyptus leaf extracts for treatment of eutrophic wastewater. *Sci. Total Environ.* 466–467, 210–213.

Wang, Z., Fang, C., Megharaj, M. 2014a. Characterization of iron-polyphenol nanoparticles synthesized by three plant extracts and their fenton oxidation of azo dye. *ACS Sustain. Chem. Eng.* 2, 1022–1025.

Wei, Y., Fang, Z., Zheng, L., Pokeung, E. 2017. Biosynthesized iron nanoparticles in aqueous extracts of *Eichhornia crassipes* and its mechanism in the hexavalent chromium removal. *Appl. Surf. Sci.* 399, 322–329.

Weng, X., Guo, M., Luo, F., Chen, Z. 2017. One-step green synthesis of bimetallic Fe/Ni nanoparticles by eucalyptus leaf extract: Biomolecules identification, characterization and catalytic activity. *Chem. Eng. J.* 308, 904–911.

Xiao, L., Mertens, M., Wortmann, L., Kremer, S., Valldor, M., Lammers, T., Kiessling, F., Mathur, S. 2015. Enhanced in vitro and in vivo cellular imaging with green tea coated water-soluble iron oxide nanocrystals. *ACS Appl. Mat. Int.* 7, 6530–6540.

Xiao, Z., Zhang, H., Xu, Y., Yuan, M., Jing, X., Huang, J., Li, Q., Sun, D. 2017. Ultra-efficient removal of chromium from aqueous medium by biogenic iron based nanoparticles. *Sep. Purif. Technol.* 174, 466–473.

Yew, Y.P., Shameli, K., Miyake, M., Kuwano, N., Khairudin, N.B.B.A., Mohamad, S.E.B., Lee, K.X. 2016. Green synthesis of magnetite (Fe_3O_4) nanoparticles using seaweed (*Kappaphycus alvarezii*) extract. *Nanoscale Res. Lett.* 11, 1–7.

Zambre, A., Upendran, A., Shukla, R., Chanda, N., Katti, K.K., Cutler, C., Kannan, R., Katti, K.V. 2012. Green nanotechnology – A sustainable approach in the nanorevolution. *Sustainable Preparation of Metal Nanoparticles* (pp. 144–156).

Sr[illegible] A.[illegible], Sudhir, A., C[illegible] K.[illegible] S[illegible] [illegible] [illegible] [illegible]tion of [illegible] nanoparticles [illegible] using the leaf extract [illegible] [illegible] [illegible] [illegible] and [illegible] [illegible] 2018 [illegible]

V[illegible]eswar[illegible], K[illegible], S[illegible], C. V[illegible], [illegible], K 2020. [illegible] [illegible] [illegible] [illegible] [illegible] magnetite nanoparticles using [illegible] [illegible] [illegible] [illegible] [illegible] [illegible]

Wang, [illegible] [illegible] Z[illegible] M[illegible] Z[illegible] [illegible] [illegible] [illegible] [illegible] leaf [illegible] [illegible] [illegible] [illegible] [illegible] 2016 [illegible]

Wang, Z[illegible] [illegible] [illegible] [illegible] [illegible] a[illegible] [illegible] [illegible] [illegible] [illegible] and [illegible] [illegible] [illegible] [illegible] [illegible] [illegible] [illegible]

[illegible], Fang [illegible], Zhang [illegible] [illegible] [illegible] [illegible] [illegible] [illegible] [illegible] [illegible] [illegible] and [illegible] in the [illegible] [illegible] [illegible] [illegible] [illegible] 32

Wang, X., [illegible] [illegible] [illegible] [illegible] [illegible] [illegible] [illegible] of [illegible] [illegible] [illegible] [illegible] [illegible] [illegible] [illegible] [illegible] [illegible] [illegible] [illegible]

[illegible], Martens, M., [illegible] [illegible] [illegible] [illegible] [illegible] [illegible] [illegible] [illegible] [illegible] [illegible] [illegible] [illegible] [illegible] [illegible] [illegible]

X[illegible] [illegible] [illegible] [illegible] [illegible] [illegible] [illegible] [illegible] [illegible] [illegible] [illegible]

[illegible] [illegible] [illegible] [illegible] [illegible] [illegible] [illegible] [illegible] [illegible]

Z[illegible] [illegible] [illegible] [illegible] [illegible] [illegible] [illegible] [illegible] [illegible] [illegible] [illegible]

5 Synthesis Utilizing Plant Wastes

5.1 FRUIT PEEL

Fruit peel/skin protects the fruit's flesh from the environment and insects. Fruit and vegetable peels have been used as natural fertilizer which are otherwise considered as waste in the society. However, such agro-waste was utilized as natural source and utilized for nanoparticle synthesis. Intensive research work has been performed utilizing fruit peel extract as source for NPs synthesis. Venkateswarlu and Yoon (2015a, 2015b) utilized *Ananas comosus* (pineapple) and *Citrullus lanatus* (watermelon) fruit peel extract for synthesizing Fe_3O_4-NPs. Such fruit peel extracts both generated spherical shaped Fe_3O_4-NPs with an average size ~17 nm. The pineapple peel extract mediated magnetic Fe_3O_4-NPs possessed saturation magnetization of 21.7 emu/g. Batch adsorption studies revealed that for 60 mg/L Cd(II) solution about 0.1 mg/L adsorbent dosage resulted in 96.1% removal at pH = 6. A maximum adsorption capacity of 49.1 mg/g was obtained. Similarly, for watermelon peel extract mediated Fe_3O_4-NPs were found to have magnetic property with 28.4 emu/g of saturation magnetization. The prepared magnetic iron oxide nanoparticle was utilized for Mercury (II) removal. Experimental batch study confirmed about 97.8% removal of Hg (II) for 60 mg/L stock solution, when treated with 0.1 mg/L dosage at pH = 7. A maximum adsorption capacity of 52.1 mg/g was obtained for watermelon peel extract mediated Fe_3O_4-NPs. The surface modified Fe_3O_4-NPs with various ligands was found to be more effective in heavy metal removal when compared with chemical synthesized iron oxide NPs. Such prepared nanoparticles had the advantage of easily being separated from large volume samples with the help of external magnetic field. Besides, easy recyclability without significantly degrading its removal efficiency are the brighter side of such nanoparticles.

Wei et al. (2016) utilized *Citrus maxima* peels as reducing agent for synthesizing iron oxide nanoparticles as shown in Figure 5.1, which were further utilized for removing Cr(VI) from aqueous solution. Such usage of fruit peel served for resource utilization along with waste minimization, which otherwise would have been discarded as waste for landfill. The extract was prepared by putting the peels with ultrapure water at high temperature for 80 min. The obtained extract was then centrifuged and membrane filtered. The obtained iron oxide NPs were spherical in shape with average diameter of 93.8 nm. Batch experiments were conducted using 100 mg/L of Cr(VI) solution using 2 mL of nanoparticle. Results conveyed that 99.29% Cr(VI) removal were obtained the end of 90 min.

DOI: 10.1201/9781003243632-5

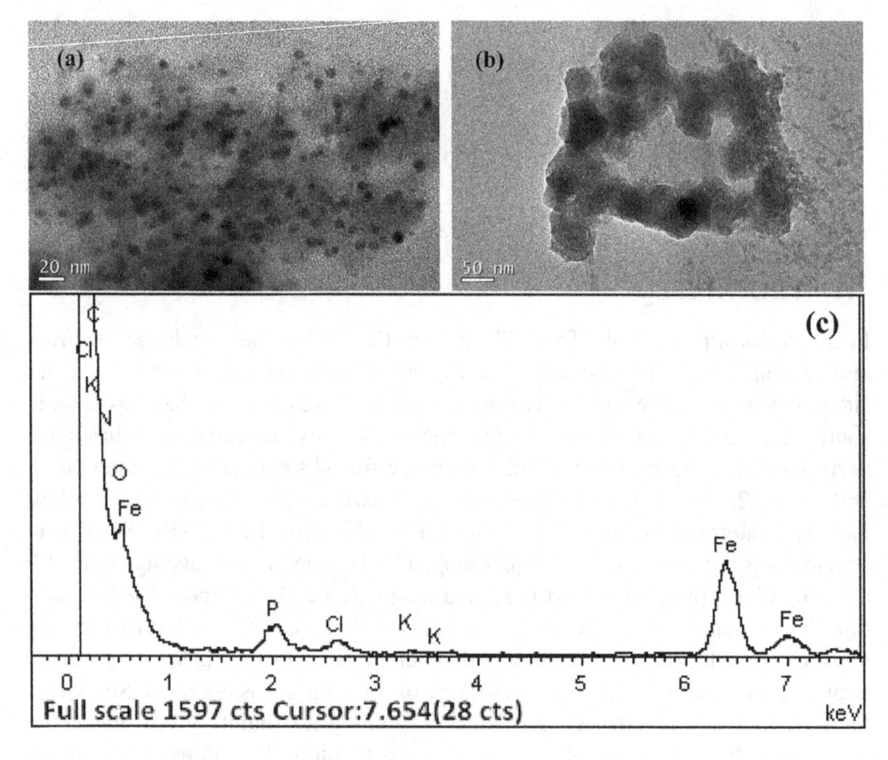

FIGURE 5.1 (a, b) TEM images and, (c) EDS spectrum of the prepared *Citrus maxima* peel-Fe-NPs [Reproduced with permission from Wei et al. (2016) © Elsevier].

Similarly Ehrampoush et al. (2015) synthesized magnetic iron oxide nanoparticles utilizing tangerine peel extract as depicted in Figure 5.2. The synthesized nanoparticles were employed for Cd(II) removal from aqueous solution. Morphological investigated revealed spherical-shaped nanoparticles with average diameter of 50 nm. The highest removal of 89.6% was observed at pH 4 for 20 mg/L Cd(II) concentration with 0.4 g/L adsorbent dosage, at the end of 90 min.

5.2 GUM

On exploring various new raw materials as a source for green synthesis, Horst et al. (2017) carried out successful research utilizing gum Arabic (GA) to produce Fe_3O_4-NPs, and further investigated the mechanism of iron oxide NPs formation. Eventually the experimental study suggests that mainly the gum polysaccharides and iron oxide nucleus went through electrostatic and/or hydrophobic interactions for iron complex formation. Figure 5.3 shows the TEM images of Arabic gum and Gum Arabic magnetite nanoparticles dispersed in various solvents.

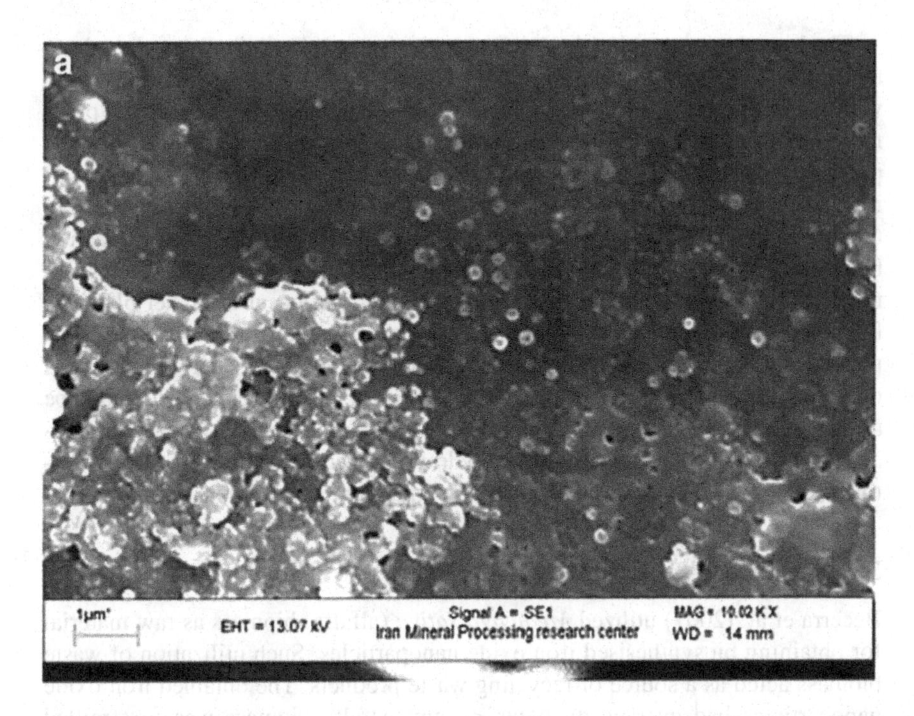

FIGURE 5.2 SEM characterization of synthesized iron oxide nanoparticles [Reproduced from Ehrampoush et al. (2015) © Springer Nature].

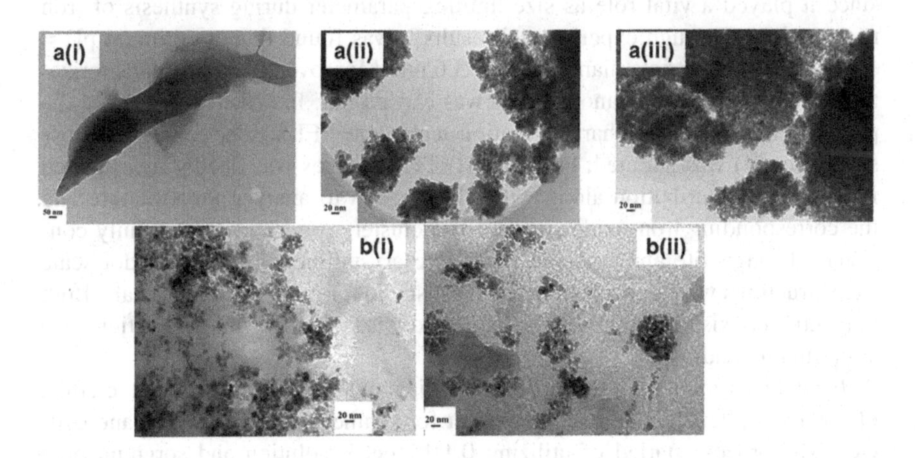

FIGURE 5.3 TEM images of Arabic gum and GA-Mag (Gum Arabic magnetite) nanoparticles: (a) ethanol/water dispersion: (i) GA, (ii) GA-Mag-a (a denotes: GA added in solid state) and (iii) GA-Mag-b (b denotes: GA added as solution); (b) water media: (i) GA-Mag-a, (ii) Ga-Mag-b [Reproduced with permission from Horst et al. (2017) © Elsevier].

The biopolymer template bridging over the iron oxide nucleus was thought to be another route for complex formation. The synthesis process, happens to begin with a acidic nature due to the presence of iron salt precursor along with the polymer moetis of GA. Iron oxide nanoparticles starts precipitation on increasing the pH with addition of NH_4OH. Opposite charge between the interating molecules of iron oxide and GA prevailed at this situation. Electrostatic interaction were thought to prevail which was better confirmed by FTIR data, showing interaction between carboxylic group and hydroxyl group of GA and iron oxide NPs respectively via hydrogen bonding. Steric interactions might occur at higher pH since both the iron oxide NPs and polymer had negative charge over their surface, further GA binding over the iron oxide NPs were resulted due to steric interactions occurring in such condition. The charged functional groups were considered to be arranged in such a way that it remains exposed over the surface. Hence the interaction between GA-Fe_3O_4 resulted in hydrophilic nature, since the surface negatively charged groups showed electrostatic repulsion between nanoparticles.

5.3 OTHER PLANT MATERIALS AND WASTES

Becerra et al. (2007) utilized *Medicago sativa* (alfalfa) biomass as raw material for obtaining biosynthesised iron oxide nanoparticles. Such utilization of waste biomass acted as a source of recycling waste products. The obtained iron oxide nanoparticles had average diameter <5 nm. Alfalfa biomass was first milled down to powder form and was then introduced to the ferrous ammonium sulphate salt solution. In this study, the effect of pH condition was determined, since it played a vital role as size-limiting parameter during synthesis of iron nanoparticle. Through experimental results it was found that at optimum pH = 10 average nanoparticle diameters was 3.6 nm. Moreover, detailed characterization of the synthesized nanoparticles was carried out. In order to determine the presence of Fe and O, characterization technique EELS (electron energy loss spectroscopy) was adopted; however, HRTEM images was deeply investigated for atomistic distribution along with FFT spectra to analyze the structure and the corresponding iron oxide species. The clusters produced were mainly consisted of magnetite and wuestite types (fcc crystalline). In quantum dot scale such structures were considered to be most stable for most of the materials. Both magnetite coexisted along with wuestite like clusters with smaller particle size at optimum conditions.

In study sorghum bran was utilized by Njagi et al. (2011) to explore the effect of abundant phenolic compounds present for synthesizing iron oxide nanoparticles. The process started of utilizing 0.1 M $FeCl_3$ solution and sorghum bran extract in a volume ratio of 2:1 for 1 h. Amorphous nature of the synthesized iron oxide nanoparticles were noticed with average spherical particle diameter of 40–50 nm as shown in Figure 5.4. The synthesized iron oxide nanoparticles were utilized for catalytic degradation of bromomethyl blue in presence of H_2O_2. A maximum degradation of 90% was achieved for 500 mg/L bromomethyl blue solution with 0.66 mM iron nanoparticles dosage in presence of 2% H_2O_2 within 30 min.

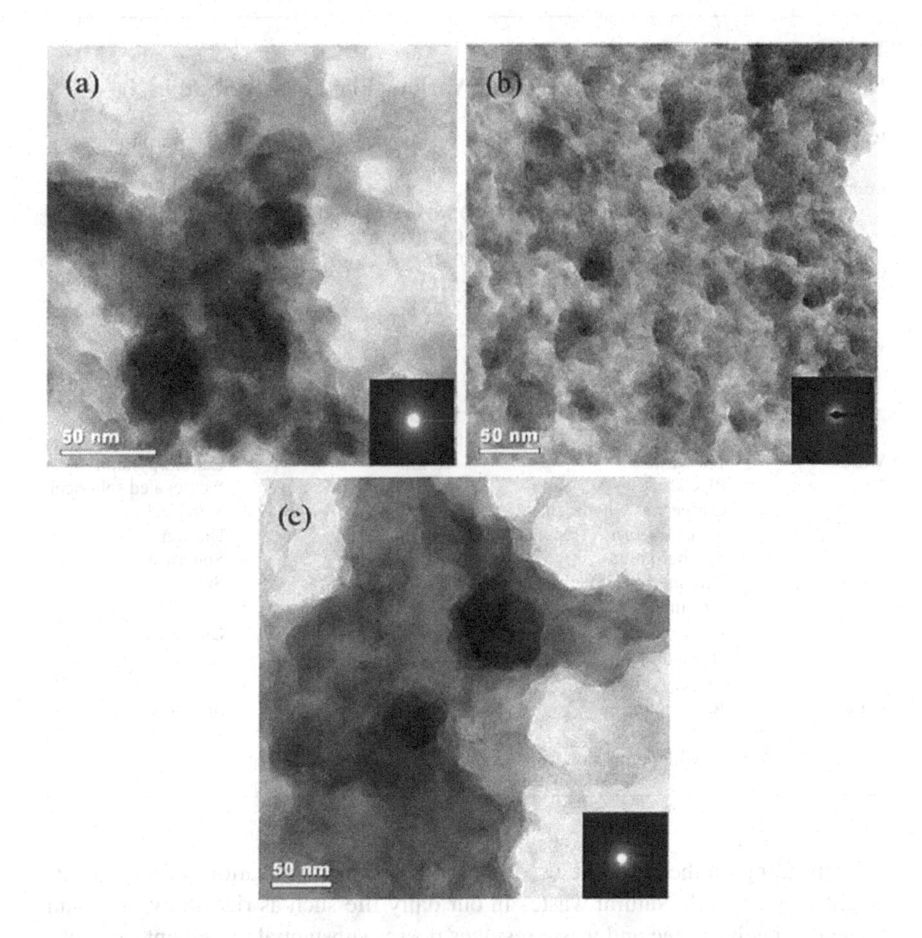

FIGURE 5.4 TEM images of iron nanoparticles synthesized using sorghum bran extracted at (a) 25, (b) 50, and (c) 80°C [Reproduced from Njagi et al. (2011) © American Chemical Society].

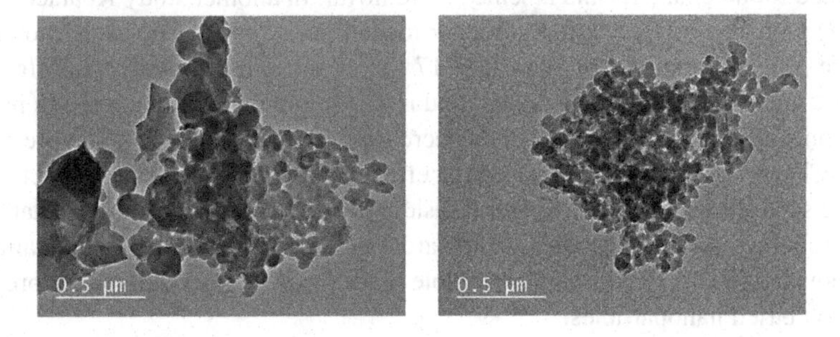

FIGURE 5.5 TEM images of Fe_3O_4-loaded coffee waste hydrochar [Reproduced with permission from Khataee et al. (2017) © Elsevier].

TABLE 5.1
Iron Nanoparticles Synthesized from Various Plant Waste Sources

Part	Name of Source	Average Size	Morphology
Fruit peel	Plantain peel	30–50 nm	Spherical
	Punica Granatum	D: 40 nm, L: >200 nm	Rod shaped
	Rambutan	100–200 nm	Agglomerated spinel
	Ananas comosus	10–16 nm	Aggregated spherical
	Citrullus lanatus	<17 nm	Aggregated spherical
	Citrus aurantium	17–25 nm	Slightly elongated
	Punica granatum		Slightly rod-shaped
	Malus domestica		Spherical
	Citrus limon		Spherical
	Plantain peel	<50 nm	Spherical
	Banana peel	10–25 nm	Agglomerated
Waste	Tea waste	5–25 nm	Cuboid/pyramid
	Rice straw	9.9 ± 2.4 nm	Aggregated spherical
	Coffee waste hydrochar	10–40 nm	Spherical
	Acacia mearnsii (biochar)	18–35 nm	Uneven
	Sorghum bran	40–50 nm	Spherical
	Cynometra ramiflora (Fruit extract waste)	58.5 nm	Spherical
	Cocos nucifera L. Chandrakalpa (Husk extract)	10–100 nm	Clustered
Gum	Arabic gum	70–80 nm	Spherical

Source: [Reproduced with permission from Mondal et al. (2020) © Elsevier].

Depending on the substrate used for synthesis, the applications of Fe_3O_4-NPs might vary as well. Natural wastes in our daily life such as rice straw, fruit and vegetable peels, coffee and waste residues posses substantial ingredient to be utilized for green synthesis of nanoparticles due to the abundance of polyphenolic content they store. Lunge et al. (2014) utilized tea residues for synthesizing Fe_3O_4-NPs and further employed it for Arsenic removal studies. Experimental studies revealed that maximum adsorption capacity of 188.69 mg/g and 153.8 mg/g were obtained for arsenic (III) and arsenic (V) removal. In another study Khataee et al. (2017) prepared coffee waste hydrochar loaded Fe_3O_4 nanoparticles as shown in Figure 5.5, and applied for Acid Red 17 (azo dye) removal study. An effective surface area 34.7 m^2/g was obtained, and it was found that with increase in initial dye concentration removal efficiency decreases from 100% to 74%. The presence of NaCl and Na_2SO_4 salts decreased the efficiency to 91%, and 85% respectively. Hence such studies prove that natural residuals provide excellent surface stabilizing properties and can be utilized towards various environmental applications. Moreover Table 5.1 enlists all the possible plant waste sources utilized for preparing iron based nanoparticles.

REFERENCES

Becerra, R.H., Zorrilla, C., Ascencio, J.A. 2007. Production of iron oxide nanoparticles by a biosynthesis method: An environmentally friendly route. *J. Phys. Chem.* 111, 16147–16153.

Ehrampoush, M.H., Miria, M., Salmani, M.H., Mahvi, A.H. 2015. Cadmium removal from aqueous solution by green synthesis iron oxide nanoparticles with tangerine peel extract. *J. Environ. Health Sci. Eng.* 13 (1), 84–90.

Horst, M.F., Coral, D.F., van Raap, M.B.F., Alvarez, M., Lassalle, V. 2017. Hybrid nanomaterials based on gum Arabic and magnetite for hyperthermia treatments. *Mater. Sci. Eng. C* 74, 443–450.

Khataee, A., Kayan, B., Kalderis, D., Karimi, A., Akay, S., Konsolakis, M. 2017. Ultrasound-assisted removal of Acid Red 17 using nanosized Fe_3O_4-loaded coffee waste hydrochar. *Ultrason. Sonochem.* 35, 72–80.

Lunge, S., Singh, S., Sinha, A. 2014. Magnetic iron oxide (Fe_3O_4) nanoparticles from tea waste for arsenic removal. *J. Magn. Magn. Mater.* 356, 21–31.

Njagi, E.C., Huang, H., Stafford, L., Genuino, H., Galindo, H.M., Collins, J.B., Hoag, G.E., Suib, S.L. 2011. Biosynthesis of iron and silver nanoparticles at room temperature using aqueous Sorghum bran extracts. *Langmuir* 27, 264–271.

Venkateswarlu, S., Yoon, M. 2015a. Surfactant-free green synthesis of Fe_3O_4 nanoparticles capped with 3,4-dihydroxyphenethylcar-bamodithioate: Stable recyclable magnetic nanoparticles for the rapid and efficient removal of Hg(II) ions from water. *Dalton Trans.* 44, 18427–18437.

Venkateswarlu, S., Yoon, M. 2015b. Rapid removal of cadmium ions using green-synthesized Fe_3O_4 nanoparticles capped with diethyl-4-(4amino-5-mercapto-4H-1,2,4 triazol-3-yl) phenyl phosphonate. *RSC Adv.* 5, 65444–65453.

Wei, Y., Fang, Z., Zheng, L., Tan, L., Tsang, E.P. 2016. Green synthesis of Fe nanoparticles using *Citrus maxima* peels aqueous extracts. *Mater. Lett.* 185, 384–386.

REFERENCES

[illegible], R.H., Comilla, C. A., [illegible] (20[illegible]). [illegible] by a biosynthesis [illegible] environmentally [illegible] [illegible]

[illegible], M.H., [illegible] M. [illegible] C[illegible] removal from [illegible] [illegible] [illegible] with [illegible]

[illegible] A. Hyb[illegible] materials [illegible] hy[illegible] [illegible] 450.

[illegible]

[illegible] S. S[illegible] [illegible] [illegible] waste [illegible] [illegible] Mat[illegible]

[illegible] M. [illegible] [illegible] [illegible] and [illegible] [illegible] 764–771.

[illegible]

[illegible]

[illegible] 1[illegible] [illegible] 1987.

6 Other Modes of Green Synthesis

6.1 MICROWAVE-ASSISTED SYNTHESIS

Recently, a microwave-assisted method has been widely applied in chemical reactions and synthesis of nanomaterials (Chen et al., 2004; Xu et al., 2007). Research has shown the method to be an attractive choice to promote reactions as it is energy efficient heating compared to conventional heat conduction methods (such as an oil bath) due to the direct heating of the reaction mixture (Nüchter et al., 2004). By conventional methods, the vessel is heated and this then transfers the heat by convection. Microwave heating is more efficient in terms of the energy used, produces higher temperature homogeneity, and is considerably more rapid than conventional heat sources.

Microwave heating is a transfer of electromagnetic energy to thermal energy, and is an energy conversion phenomenon rather than heat transfer. The electric component of an electromagnetic field causes heating by two main mechanisms, dipolar polarization mechanism and conduction mechanism (Stass et al., 2000). In polarization mechanism, for a substance to generate heat when irradiated with microwaves it must possess a dipole moment, as in water molecule. A dipole is sensitive to external electric fields and will attempt to align itself with the field by rotation. The applied field provides the energy for this rotation (Lidström et al., 2001). As molecules vainly attempt to follow the field, they collide with one another and heating in the sample is observed. High- and low-frequency radiation does not give rise to efficient heating: in the first case, the field oscillates too quickly for the molecules to respond; in the second case, the molecules follow the field so well that there is no random motion generated (Whittaker, 2004).

If two samples containing distilled water and tap water, respectively, are heated in a single mode microwave cavity at a fixed radiation power and for a fixed time, the final temperature will be higher in the tap water sample. This phenomenon is due to the second major interactions of the electric field component with the sample, the conduction mechanism. The conduction mechanism is a much stronger interaction than the dipolar mechanism with regard to the heat-generating capacity (Lidström et al., 2001). In this case, any mobile charge carriers (electrons, ions, etc.) move relatively easily through the material under the influence of the microwave's electric field. These induced currents will cause heating in the sample due to any electrical resistance. If the sample is a metallic conductor, most of the microwave energy is reflected with relatively little energy penetrating beyond a few microns into the surface. However, the surface electrons respond to the field

DOI: 10.1201/9781003243632-6

in less than 10–12 s, and colossal surface potentials (of the order of hundreds of kV m^{-1}) may be induced. Conduction heating itself can be demonstrated quite easily in a domestic microwave oven using materials such as copper oxide or carbon. Alternatively, if pure water is heated in a microwave oven, where the polarization mechanism dominates, the heating rate is significantly lower than that observed when the same volume of a dilute salt solution is heated. In the latter case, both dipolar polarization and conductive mechanisms contribute to the heating effect (Whittaker, 2004).

Microwave energy is a spectrum in the same way as visible light, infrared irradiation, and UV irradiation and is delivered directly to the material through molecular interaction with the electromagnetic field. Since microwaves can penetrate the material and supply energy, heat can be generated throughout the volume of the material resulting in volumetric heating (Das et al., 2009). Additionally, the method shows acceleration in reaction rate, yield improvement, short reaction time, small particle size, narrow particle size distribution, high purity materials, and enhanced physicochemical properties (Vovchenko et al., 2004; Miyawaki et al., 2006; Li et al., 2007). The results obtained cannot be explained by the effect of the rapid heating alone, and this has led various authors to postulate the existence of "microwave effect". Historically, "specific microwave effects" have been claimed, when the outcome of a synthesis performed using microwave heating differs from the conventionally heated counterpart carried out at the same apparent temperature (Das et al., 2009).

Electromagnetic irradiation was utilized through microwave-assisted synthesis where via ionic and molecular conduction the precursor solvents and reducing agents were involved in reaction. Łuczak et al. (2016) utilized this method where heating at high temperature was employed within the sample directly. Since the requirement of energy is less, hence microwave irradiation provides rapid heating facilitating greener preparation of nanoparticles. Such process regulates uniform dispersion of nanoparticles, whereas nucleation accelerates. Preparation of nanoparticles within minutes through such heating methods with efficiency in control over size regulation, crystallinity, and dispersion are considered as crucial factors for greener synthesis techniques (Amores et al., 2016; Ortega et al., 2015). A list of various microwave-assisted iron and iron-based nanomaterials synthesized by various researchers are mentioned in Table 6.1.

Kombaiah et al. (2017) in their study showed that microwave technique was considerably more efficient and economical with respect to other thermal techniques where energy consumption and fabrication cost of nanoparticles synthesis are relatively lower. Schneider et al. (2017) described the fast synthesis of alendronic acid (a reactive biphosphate) coated magnetic nanoparticle through green microwave-assisted technique. Synthesis of uniform nanoparticles at heating temperature of 200°C was obtained having a small particle size average. The process includes iron acetylacetonate in triethylene glycol (TEG) as precursor solution and reaction was maintained at inert condition. TEG is considered as a green component due to its low isoelectric point, high viscosity, nontoxicity, and high boiling point.

TABLE 6.1
Types of Microwave-assisted Biosynthesis of Iron-based Nanoparticles

Nanoparticle	Stabilizer	NP Size	Morphology	Reference
Magnetic iron oxide (Fe_3O_4) nanoparticles	poly(sodium 4-styrenesulfonate (PSSS) and sodium polyphosphate (SPP)	12 ± 2 nm	Aggregated irregular NPs	Williams et al. (2016)
Magnetic iron oxide (Fe_3O_4 and γ-Fe_2O_3) nanoparticles	Mixture of oleic acid with oleylamine	~5 nm	Organic shell covered NPs	Lastovina et al. (2017)
MCM-41 supported iron oxide nanoparticles (FeO and Fe_2O_3)	Tetraethylorthosilicate (TEOS, 98 %) with NH_4OH solution as surfactant	~3 nm	Narrow distributed pore size	Carrillo et al. (2013)
Magnetic iron oxide nanoparticles coated with nano silica layer	1.0 mol/L sodium silicate solution	33–66 nm	Homogenous shaped	Mahmoud et al. (2016)
Iron molybdate microspheres & nanoparticles	Ethanol (for synthesis of NPs) act as dispersant	5–20 nm	Flower-shaped microspheres	Liang et al. (2016)
Magnetite nanoparticles	alendronic acid (a biphosphate)	5.9 ± 1.4 nm	Uniform NPs	Schneider et al. (2017)
Iron oxide nanoparticles	Trisodium citrate dihydrate (Na_3Cit)	5–50 nm	Crystalline, irregular shaped	Gonzalez-Moragas et al. (2015)
Cobalt ferrite ($CoFe_2O_4$) nanoparticles	Okra (*A. esculentus*) plant extract used as reductant	5–15 nm	Spherical	Kombaiah et al. (2017)
Zinc ferrite ($ZnFe_2O_4$) nanoparticles	Okra (*A. esculentus*) plant extract used as reductant	7.2 ± 1.3 nm	Agglomerated spherical	Kombaiah et al. (2017)
Superparamagnetic iron oxide nanoparticles (SPIONs)	trimethylammonium hydroxide (TMAOH) solution and sodium citrate used as anionic stabilizers	19–44 nm	Crystalline round lobular	Carenza et al. (2014)
Hematite (α-Fe_2O_3) nanorods	$(NH_4)_3PO_4$	Nanorod length ~150–170 nm, diameter ~60 nm	Rod-shaped	Ahmed et al. (2014)

Source: [Reproduced with permission from Mondal et al. (2020) © Elsevier].

Low yield of nanoparticle generation has been one of the main drawbacks for its large scale application due to limited reaction vessel dimension in microwave assisted synthesis. In order to overcome such limitation Gonzalez-Moragas et al. (2015) utilized a multimode microwave unit to scale-up the synthesis of multigram iron oxide nanoparticles. First, utilizing iron (III) acetylacetonate ($Fe(acac)_3$) in anhydrous benzyl alcohol as precursor the nanoparticle synthesis at laboratory-scale was carried out. Increment of temperature was performed from room temperature to 180°C within 20 minutes and then at 300 W the process was cooled over three minutes. Average particle size analyzed through DLS and TEM characterization was obtained at around 3.8 ± 0.8 nm. During scale-up, the process was carried out at a higher microwave power of 500 W under optimized conditions and heating temperature was kept to maximum at 200°C. The reaction time was extended for the process which resulted in tenfold times scale-up. Utilizing the scale-up technique an excellent yield >80 % was achieved.

A magnetically recoverable nanocatalyst was successfully fabricated by Li et al. (2016) utilizing the technique of grafting silver nanoparticles onto a carboxymethyl cellulose support. The process was carried out under microwave irradiation which resulted in impregnating magnetic iron oxide nanoparticles into the support and further utilized for degradation of carbonyls to carboxylic acids by catalyzed hydrogenation in water.

Similarly, synthesis of multifunctional magnetite nanoparticles impregnated on polymer support was reported by Williams et al. (2016) via a microwave heating method. The process was carried out at 150°C for 20 minutes. In this process $FeCl_3.6H_2O$ and $FeCl_2.4H_2O$ were used as precursors and single step synthesized aggregated, high crystalline nanoparticles were generated with better aqueous stability. Moreover, a study showed that a two-step microwave heating process could be utilized for the synthesis of magnetite and maghemite nanoparticles. In such process iron (III) acetylacetonate was utilized as precursor, dissolved in a mixture of oleic acid and oleylamine acting as stabilizing agent. An average particle size of 5nm was obtained when different ratios of both stabilizers were utilized for nanoparticle synthesis. Such process was operated by heating the different stabilizer obtained mixture at 120°C for 1 hour and another at 185°C for 1.5 hours respectively with continuous stirring. Among the stabilizers oleic acid was found to have greater effect in minimizing agglomeration of the nanoparticles formed (Lastovina et al., 2017).

Grindi et al. (2018) reported the synthesis of nanoparticles utilizing the combination of microwave with other thermal techniques like, hydrothermal and solvothermal methods. Liang et al. (2016) in his study stated the advantage of such process in preparing high crystalline nanoparticles due to extremely fast heating rate at high temperature. Moreover, such methods are applicable for both polar and nonpolar solvents. Grindi et al. (2018) synthesized one-pot strontium hexaferrite nanoparticles, utilizing iron (III) nitrate nonahydrate and strontium nitrate as precursors utilizing microwave-assisted hydrothermal technique. The SEM images of the synthesized NPs are shown in Figure 6.1. The process was carried out at an optimum irradiation condition which was achieved at 200°C with heating rate 25°C/min maintained for one hour. Due to the high heating rate of microwave-assisted irradiation in shorter time, the crystallization of prepared hematite was avoided, thus microwave-assisted synthesis proves to be beneficial in comparison with classical hydrothermal process.

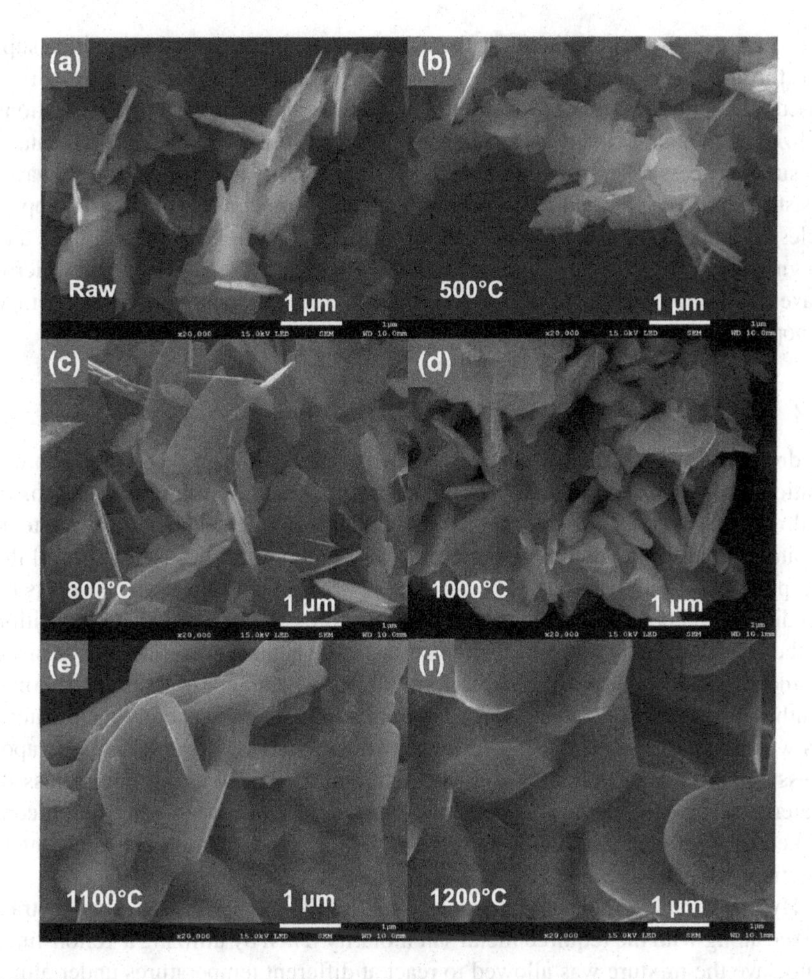

FIGURE 6.1 SEM images of M-$SrFe_{12}O_{19}$ powders as-prod (a) and annealed for 1 h at 500°C (b), 800°C (c), 1000°C (d), 1100°C (e) and 1200°C (f) [Reproduced with permission from Grindi et al. (2018) © Elsevier].

Mahmoud et al. (2016) reported that through microwave-assisted techniques it was possible to synthesize surface impregnated nanoparticles over or within polymeric supports which enhances the dispersion and stability of nanoparticles. Such enhanced property was hard to achieve for unsupported nanoparticles. Moreover, synthesis of MCM-41 silica supported surface loaded with iron oxide nanoparticles was reported by Carrillo et al. (2013). Tetraethylorthosilicate as precursor and ammonium hydroxide solution as surfactant were utilized for synthesizing silica with helical mesoporous and well-ordered sphere structures having core shell design. $FeCl_2.4H_2O$ precursor-based iron oxide nanoparticles were coated over silica support through the microwave-assisted technique by heating the mixture containing precursor and support at 200 W for 15 minutes. An average pore diameter of 3nm was obtained for the nanoparticles incorporated over silica support through XRD analysis. Kombaiah et al. (2017) biosynthesized ceramic iron-based nanoparticles assisted by microwave-based technique – such reactions utilized okra plant extracts as reducing agents. The study compares the advantage of such process over other conventional heating methods. Both microwave-assisted, and conventional techniques procured agglomerated, spherical nanoparticles of cobalt and zinc ferrite nanoparticles respectively. Better optical and magnetic property oriented nanoparticles were synthesized through the microwave heating method as compared to conventional methods with lower average nanoparticle diameter.

6.2 HYDROTHERMAL SYNTHESIS

Hydrothermal synthesis is one of the most commonly used methods for preparation of nanomaterials. It is basically a solution reaction-based approach. In hydrothermal synthesis, the formation of nanomaterials can happen in a wide temperature range from room temperature to very high temperatures. To control the morphology of the materials to be prepared, either low-pressure or high-pressure conditions can be used depending on the vapor pressure of the main composition in the reaction. Many types of nanomaterials have been successfully synthesized by the use of this approach. There are significant advantages of the hydrothermal synthesis method over others. Hydrothermal synthesis can generate nanomaterials which are not stable at elevated temperatures. Nanomaterials with high vapor pressures can be produced by the hydrothermal method with minimum loss of materials. The compositions of nanomaterials to be synthesized can be well controlled in hydrothermal synthesis through liquid phase or multiphase chemical reactions.

Hydrothermal synthesis method involves utilizing the desired plant extract mixed along with the required metal salt molarity. Then by utilizing a Teflon-lined autoclave the mixture was allowed to react at different temperatures under atmospheric pressure for a fixed interval of time. Such process is beneficial since it requires a lower temperature to produce crystalline materials from precursor than calcination.

Ahmmad et al. (2013) using green tea (*Camellia sinensis*) leaf extract, through hydrothermal synthesis method successfully prepared highly pure hematite α-Fe_2O_3 nanoparticles. TEM characterization indicated spherical nanoparticles with an average diameter 60 nm and highly porous in nature. The synthesized hematite nanoparticles were found to have higher surface area (22.50 m^2/g) and enhanced photocatalytic activity as compared to commercially available hematite nanoparticles. Surface area was nearly four times higher, whereas the photocatalytic activity was found to be nearly two times higher than commercially available ones.

A plausible formation mechanism of porous iron oxide is schematically presented in Figure 6.2, assuming that epigallocatechin gallate (EGCG) or gallocatechin gallate (GCG) works as surfactant, which might be deprotonated at elevated

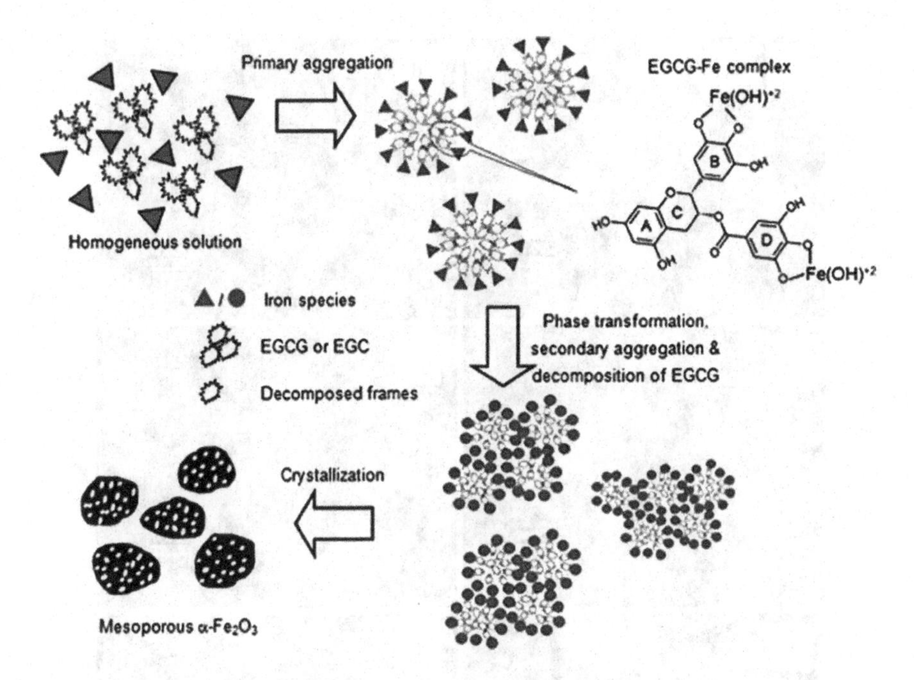

FIGURE 6.2 Schematic illustration for the formation-mechanism of α-Fe_2O_3 nanoparticles [Reproduced with permission from Ahmmad et al. 2013 © Elsevier].

temperature and pressure during the hydrothermal process. The initial octahedral aqua complex of iron (III), $Fe(H_2O)_6$, which is formed in water instantaneously decomposes into several soluble low molecular weight species (such as $[Fe(OH_2)(OH)]^{2+}$, $[Fe(OH_2)_4(OH)_2]^{+}$) via deprotonation of coordinated water molecule (Khaleel, 2004). The hydrolyzed iron species could form a complex with B and D ring of deprotonated EGCG (Figure 6.2) at halfway of the hydrothermal process (comparatively at low temperature and pressure) (Ryan and Hynes, 2007; El-Ayaan et al., 1998). As the other side of EGCG (A-ring side) is comparatively hydrophobic in nature, a tendency of aggregation might occur forming a structure shown in Figure 6.2. With the increase in the temperature a phase transformation of iron species $[Fe(OH)_2]^{+}$ might occur forming primary particles (Fe_2O_3) of several nanometer. These primary particles have high surface energy and aggregate quickly to minimize their surface energy (Wan et al., 2011). The EGCG or GCG molecules might be decomposed at high temperature and the decomposed products are dispersed into water leaving pore in the particles.

Moreover, Fe_3O_4 nanoparticles were synthesized using aloe vera plant extract by hydrothermal method (Phumying et al., 2012). In the process ferric acetylacetonate ($Fe(C_5H_8O_2)_3$) acted as precursor. Crystalline spherical nanoparticles with average diameter 6–30 nm were obtained as revealed through TEM characterization. Detailed study through XRD and HR-TEM conveys that the Fe_3O_4 nanoparticles had inverse cubic spinel morphology with absence of phase impurities.

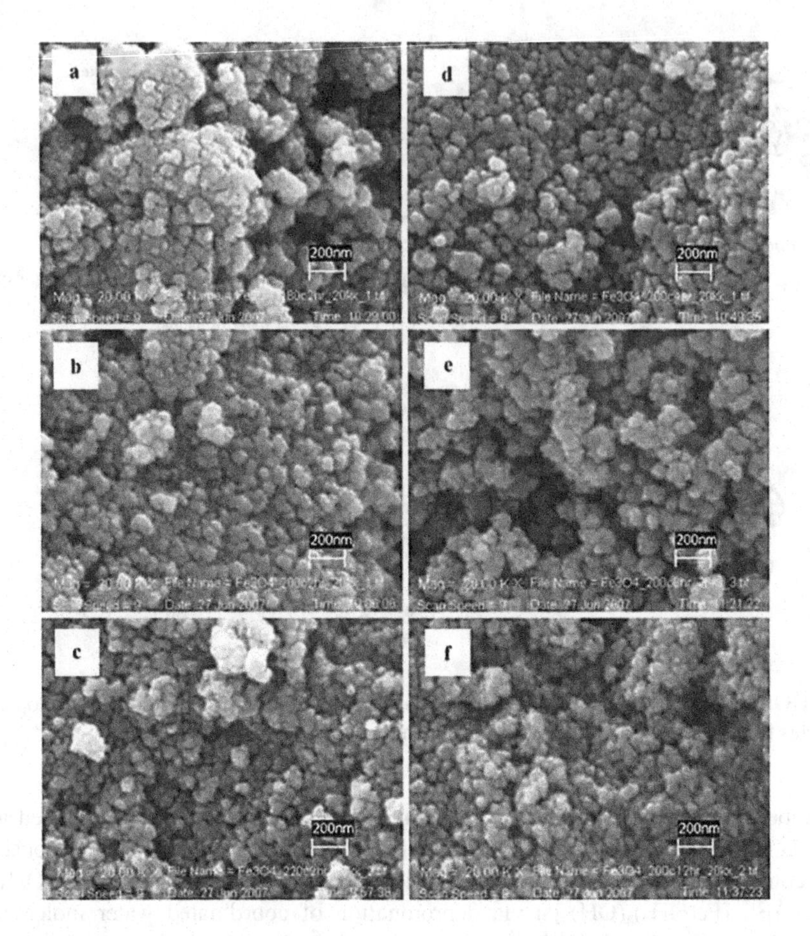

FIGURE 6.3 SEM micrographs of Fe_3O_4 nanoparticles prepared by a novel hydrothermal technique using aloe vera plant-extracted solution. The hydrothermal time and temperature used are: (a) 180°C/2 h; (b) 200°C/2 h; (c) 220°C/2 h; (d) 200°C/4 h; (e) 200°C/8 h; and (f) 200°C/12 h [Reproduced with permission from Phumying et al. (2012) © Elsevier].

Superparamagnetic nature was confirmed based on coercivity experiment. The study shows that magnetic nanoparticles with enhanced crystallinity and saturation magnetization could be procured by increasing the reaction temperature and time. The morphology of the Fe_3O_4 samples prepared with different reaction temperatures and times was investigated by SEM. It is clearly seen from the SEM micrographs (Figure 6.3a–f) that the morphology of all of the materials are agglomerated nanoparticles with sizes of ~25–50 nm. It is noted that the particles sizes estimated by SEM are larger than those obtained from X-ray line broadening. This is due to agglomeration of the nanoparticles in the samples.

REFERENCES

Ahmmad, B., Leonard, K., Shariful Islam, M., Kurawaki, J., Muruganandham, M., Ohkubo, T., Kuroda, Y. 2013. Green synthesis of mesoporous hematite (α-Fe_2O_3) nanoparticles and their photocatalytic activity. *Adv. Powder Technol.* 24, 160–167.

Ahmed, F., Arshi, N., Anwar, M.S., Danish, R., Koo, B.H. 2014. Quantum-confinement induced enhancement in photocatalytic properties of iron oxide nanoparticles prepared by ionic liquid. *Ceram. Int.* 40 (10), 15743–15751.

Amores, M., Ashton, T.E., Baker, P.J., Cussen, J., Corr, S.A. 2016. Fast microwave-assisted synthesis of Li-stuffed garnets and insights into Li diffusion from muon spin spectroscopy. *J. Mater. Chem. A Mater. Energy Sustain.* 4, 1729–1736.

Carrillo, A.I., Serrano, E., Luque, R., Garcia-Martinez, J. 2013. Microwave-assisted catalysis by iron oxide nanoparticles on MCM-41: Effect of the support morphology. *Appl. Catal. A Gen.* 453, 383–390.

Carenza, E., Barcelo, V., Morancho, A., Montaner, J., Rosell, A., Roig, A. 2014. Rapid synthesis of water-dispersible superparamagnetic iron oxide nanoparticles by a microwave-assisted route for safe labeling of endothelial progenitor cells. *Acta Biomater.* 10 (8), 3775–3785.

Chen, W. X., Lee, J. Y., Liu, Z. 2004. Preparation of Pt and PtRu nanoparticles supported on carbon nanotubes by microwave assisted heating polyol process. *Mat. Let.* 58 (25), 3166–3169.

Das, S., Mukhopadhyay, A. K., Datta, S., Basu, D. 2009. Prospects of microwave processing: An overview. *Bulletin Mat. Sci.* 32 (1), 1–13.

El-Ayaan, U., Jameson, R.F., Linert, W. 1998. A kinetic study of the reaction between noradrenaline and iron(III): An example of parallel inner-andouter-sphere electron transfer. *J. Chem. Soc. Dalton Trans.* 1315–1319.

Gonzalez-Moragas, L., Yu, S.-M., Murillo-Cremaes, N., Laromaine, A., Roig, A. 2015. Scale-up synthesis of iron oxide nanoparticles by microwave-assisted thermal decomposition. *Chem. Eng. J.* 281, 87–95.

Grindi, B., Beji, Z., Viau, G., BenAli, A. 2018. Microwave-assisted synthesis and magnetic properties of M-$SrFe_{12}O_{19}$ nanoparticles. *J. Magn. Magn. Mater.* 449, 119–126.

Khaleel, A.A. 2004. Nanostructured Purec-Fe_2O_3 via forced precipitation in anorganic solvent. *Chem. Eur. J.* 10, 925–932.

Kombaiah, K., Vijaya, J.J., Kennedy, L.J., Bououdina, M. 2017. Optical, magnetic and structural properties of $ZnFe_2O_4$ nanoparticles synthesized by conventional and microwave assisted combustion method: A comparative investigation. *Opt. – Int. J. Light Electron. Opt.* 129, 57–68.

Lastovina, T.A., Budnyk, A.P., Soldatov, M.A., Rusalev, Y.V., Guda, A.A., Bogdan, A.S., Soldatov, A.V. 2017. Microwave-assisted synthesis of magnetic iron oxide nanoparticles in oleylamine–oleic acid solutions. *Mendeleev Commun.* 27 (5), 487–489.

Li, A.Y., Kaushik, M., Li, C.J., Moores, A. 2016. Microwave-assisted synthesis of magnetic carboxymethyl cellulose-embedded Ag-Fe_3O_4 nanocatalysts for selective carbonyl hydrogenation. *ACS Sustain. Chem. Eng.* 4 (3), 965–973.

Li, Y., Lee, E.J., Cho, S.O. 2007. Superhydrophobic coatings on curved surfaces featuring remarkable supporting force. *J. Phy. Chem. C* 111 (40), 14813–14817.

Liang, J., Zhuo, M., Guo, D., Chen, Z., Ren, W., Zhang, M., Li, Q. 2016. Green and rapid synthesis of 3D $Fe_2(MoO_4)_3$ by microwave irradiation to detect H2S gas. *Mater. Lett.* 168 (3), 171–175.

Lidström, P., Tierney, J., Wathey, B., Westman, J. 2001. Microwave assisted organic synthesis – A review. *Tetrahedron* 57 (51), 9225–9283.

Łuczak, J., Paszkiewicz, M., Krukowska, A., Malankowska, A., Zaleska-Medynska, A. 2016. Ionic liquids for nano- and microstructures preparation. Part 2: Application in synthesis. *Adv. Colloid Interface Sci.* 227, 1–52.

Mahmoud, M.E., Amira, M.F., Zaghloul, A.A., Ibrahim, G.A.A. 2016. High performance microwave-enforced solid phase extraction of heavy metals from aqueous solutions using magnetic iron oxide nanoparticles-protected-nanosilica. *Sep. Purif. Technol.* 163, 169–172.

Mondal, P., Anweshan, A., Purkait, M.K. 2020. Green synthesis and environmental application of iron-based nanomaterials and nanocomposite: A review. *Chemosphere* 259, 127509.

Miyawaki, J., Yudasaka, M., Imai, H. 2006. In vivo magnetic resonance imaging of single-walled carbon nanohorns by labelling with magnetite nanoparticles. *Adv. Mat.* 18 (8), 1010–1014.

Nüchter, M., Ondruschka, B., Bonrath, W., Gum, A. 2004. Microwave assisted synthesis – A critical technology overview. *Green Chem.* 6 (3), 128–141.

Ortega, D., Southern, P., Pankhurst, Q.A., Thanh, N.T.K. 2015. High performance multi-core iron oxide nanoparticles for magnetic hyperthermia: Microwave synthesis, and the role of core-to-core. *Nanoscale* 7, 1768–1775.

Phumying, S., Labuayai, S., Thomas, C., Amornkitbamrung, V., Swatsitang, E., Maensiri, S. 2012. Aloe vera plant-extracted solution hydrothermal synthesis and magnetic properties of magnetite (Fe_3O_4) nanoparticles. *Appl. Phys. A* 111, 1187–1193.

Ryan, P., Hynes, M.J. 2007. The kinetics and mechanisms of the complex formation and antioxidant behaviour of the polyphenols EGCg and ECG with iron (III). *J. Inorg. Biochem.* 101, 585–593.

Schneider, T., Lowa, A., Karagiozov, S., Sprenger, L., Gutierrez, L., Esposito, T., et al., 2017. Facile microwave synthesis of uniform magnetic nanoparticles with minimal sample processing. *J. Magn. Magn. Mater.* 421, 283–291.

Stass, D.V., Woodward, J.R., Timmel, C.R., Hore, P.J., McLauchlan, K.A. 2000. Radiofrequency magnetic field effects on chemical reaction yields. *Chem. Phy. Let.* 329(1–2), 15–22.

Vovchenko, L., Matzui, L., Zakharenko, M., Babich, M., Brusilovetz, A. 2004. Thermoexfoliated graphite as support for production of metal-graphite nanocomposites. *J. Phys. Chem. Solids* 65(2–3), 171–175.

Wan, L., Yan, S., Wang, X., Li, Z., Zou, Z. 2011. Solvothermal synthesis of mono-disperseiron oxides with various morphologies and their applications in removal of Cr (VI). *Crystal. Eng. Com.* 13, 2727.

Whittaker, G. 2004. Microwave chemistry. *School of Sci. Rev.* 85 (312), 87–94.

Williams, M.J., Sanchez, E., Aluri, E.R., Douglas, F.J., MacLaren, D.A., Collins, O.M., et al. 2016. Microwave-assisted synthesis of highly crystalline, multifunctional iron oxide nanocomposites for imaging applications. *RSC Adv.* 6 (87), 83520–83528.

Xu, Q.C., Lin, J.D., Li, J., Fu, X.Z., Liang, Y., Liao, D.W. 2007. Microwave-assisted synthesis of MgO-CNTs supported ruthenium catalysts for ammonia synthesis. *Cat. Com.* 8(12), 1881–1885.

7 Environmental Applications of Green Synthesized Iron Nanoparticles

7.1 DYE DEGRADATION APPLICATION

For synthesizing iron nanoparticles Hoag et al. (2009) employed green tea and utilized it for catalyzing hydrogen peroxide for bromomethyl blue degradation in aqueous solution. It was concluded from the study that with respect to Fe-EDTA and Fe-EDDS, enhanced catalytic activity was produced by such nanoscale zero-valent iron obtained by green synthesis (nZVI). With increasing dosage of green synthesized nZVI, the degradation of bromomethyl blue was found to increase due to enhanced catalyzation of hydrogen peroxide. Several studies related to dye removal applications utilizing various green synthesized iron-based nanomaterials are reported in Table 7.1.

TABLE 7.1
Green Synthesized Iron-based Nanoparticles for Dye Degradation Applications

Source	Nanoparticle	Morphology	Dye Removed	Efficiency
Pomegranate leaves	Fe_3O_4 NPs	45–60 nm, spherical	Congo red	93% removal at 60 min
Pisum sativum peels	Iron oxide magnetic NPs	20–30 nm, spherical	Methyl orange	96.2% removal at 60 min
Eucalyptus leaf extract	Fe/Ni bimetallic NPs	20–50 nm, spherical abd irregular	Methyl orange	99.6% removal at 180 min
Cupressus sempervirens leaf	nZVI	1.5 nm, nanoclusters	Methyl orange	95.8% removal at 6h
Moringa oleifera leaves	Ni/Fe_3O_4 NPs	16–20 nm, spherical	Malachite green	~91.6% removal at 25 min
Green tea leaves	Superparamagnetic Fe_3O_4 nanoparticles	10 ± 3 nm, spherical, narrow size distribution	Methylene blue	95% removal at 16 min

(Continued)

DOI: 10.1201/9781003243632-7

TABLE 7.1 (Continued)

Source	Nanoparticle	Morphology	Dye Removed	Efficiency
Pepper extract	Iron oxide/ palladium nanocomposite	~50 nm, spherical and dendrite nanostructures	Acid black acid brown	97% removal at 120 min
Tie Guanyin tea extract	nZVI	6.58 ± 0.76 nm, spherical	Bromothymol blue	>90% removal at 30 min
Cynometra ramiflora fruit extract	Fe_3O_4 NPs	58.5–78.1 nm, spherical	Methylene blue	100% removal at 110 min
Cynometra ramiflora	Fe_3O_4 NPs	>100 nm, spherical aggregates	Rhodamine B	100% removal at 15 min
Camellia sinensis	nZVI	5–15 nm, spherical crystalline	Bromothymol blue	83.3% removal at 20 min
Green tea	Fe_3O_4 NPs	40–60 nm, amorphous	Aqueous cationic and anionic dyes	100% removal at 6h
Eucalyptus tereticornis	Fe-Polyphenol NPs	40–60 nm, cubical	Azo dyes (adsorption)	Maximum adsorption capacity 1.6 g dye/g NPs
Eucalyptus tereticornis, Melaleuca nesophila, and *Rosemarinus officinalis*	Fe-Polyphenol NPs	50–80 nm, spherical	Azo dyes (degradation)	100% removal at 200 min
Grape	Fe_3O_4 NPs	15–100 nm, quasi-spherical shaped	Acid orange (Azo dues)	80% removal at 180 min
Sorghum bran	Fe NPs	4–50 nm, spherical amorphous	Bromomethyl blue	90% removal at 30 min
Oolong tea leaves	Fe NPs	40–50 nm, spherical	Malachite green	75.5% removal at 60 min
Green tea	Fe NPs	70–80 nm, spherical amorphous	Malachite green	82.47% removal at 60 min

Source: [Reproduced with permission from Mondal et al. (2020) © Elsevier].

Similarly, Njagi et al. (2011) synthesized iron nanoparticles utilizing sorghum bran extracts and studied their catalytic activity towards degradation of the dye bromothymol blue. Degradation of bromomethyl blue in the presence of iron nanoparticles and H_2O_2, dictates that the synthesized nanoparticles produced free radicals from H_2O_2 which catalyzed the degradation reaction for bromothymol blue. The rate of reaction depends on the catalysis of H_2O_2 which ultimately enhanced the degradation rate of bromothymol blue.

In another report, Shahwan et al. (2011) synthesized nZVI nanoparticles utilizing green tea extract. The as-synthesized nZVI was applied toward degradation of methylene blue (MB) and methylene orange (MO) dyes. With an adsorbent dosage of 10–200 mg/L nearly 100% removal of methylene blue (MB) and methyl orange (MO) was achieved. From experimental study it was observed that MB removal was achieved faster than MO dye. In the first five minutes, about 80% of MB was removed whereas, after 1 hour 80% removal of MO dye was attained. A complete removal of MB and MO dye was achieved after 200 minutes and 350 minutes respectively. Studies convey that iron nanoparticles synthesized through green routes are comparatively more effective than chemically reduced nanoparticles for acting as a Fenton-like catalyst.

Oolong tea extract was utilized for synthesis of iron nanoparticles by Huang et al. (2013). The nanoparticles were employed for malachite green (MG) degradation. The experimental results conveyed that the polyphenolic content in oolong tea extract served the purpose of both reducing and capping agents during the synthesis. The high polyphenolic content paved the way for lesser aggregation along with enhanced reactivity of the synthesized nanoparticles. On application toward 50 mg/L MG solution the nanoparticles resulted in 75.5% dye removal.

A similar study was carried out by Kuang et al. (2013) where three different tea extracts were used:oolong tea (OT), green tea (GT), and black tea (BT) separately for the iron nanoparticles synthesis. Furthermore, the synthesized nanoparticles were utilized as a catalyst for degradation of monochlorobenzene (MCB) for a Fenton-like oxidation. The highest polyphenol content of the chosen extracts resulted in better synthesized Fe NPs for enhanced degradation purposes. GT-based iron NPs resulted in 69% MCB degradation, whereas OT- and BT-based nanoparticles resulted in 53% and 39% MCB degradation in 180 minutes.

Wang (2013) discusses combating dye degradation synthesized nanoparticles involving iron-polyphenol complex. Eucalyptus leaves were utilized as reducing and capping agent, and the synthesized nanoparticles were employed against acid black 194 dye for adsorption-flocculation test. Adsorption capacity of the synthesized nanoparticles was found to be 1.6 g acid black 194/g of nanoparticles at 25°C.

Moreover, Wang et al. (2014a) utilized three different plant sources i.e., *E. tereticornis*, *M. nesophila* and *R. officinalis* for synthesizing Fe NPs and were utilized towards decolorization of acid black dye. The extract of *E. tereticornis* generated Fe NPs showed 100% removal of acid black dye along with 87% removal of total organic carbon (TOC). Such Fe NPs showed better catalytic properties when compared to the others. The as-synthesized nanoparticles showed lower average diameter with better dispersion property in aqueous medium.

Huang et al. (2015) reported that the variables in an experiment such as the volume ratio of Fe ions and tea extract, temperature, and pH had influencing effects during the synthesis of nanoparticles. With increase in tea extract concentration, the synthesized Fe NPs concentration decreases due to lower Fe ion concentration. Further studies revealed the effect of synthesized Fe NPs, catalytic reactivity for degradation of malachite green (MG) dye. The degradation study was found to vary along with NPs synthesized condition, pH, and moreover increased temperature increased the rate of degradation.

In another study Luo et al. (2014) investigated the removal of acid orange II dye by utilizing iron nanoparticles synthesized by grape leaf extract. The study revealed the reactivity of plant mediated Fe NPs was better than the Fe NPs synthesized from methanolic and aqueous extract of grape leaves. Hence, under various experimental conditions plant mediated Fe nanoparticles proved to be more effective for catalytic degradation of dyes. Plant mediated Fe-NPs proved to be more suitable for Fenton-like reaction compared to the conventional Fenton reaction, where it acts as a Fenton catalyst with H_2O_2. Several studies carried by Prasad et al. (2017a) report pomegranate leave extract mediated iron oxide nanoparticle synthesis. The synthesized iron oxide nanoparticles were utilized for adsorption study of Congo red dye. The experiments conducted resulted in 93% dye removal within an hour.

Moreover, further research investigated iron oxide nanoparticle synthesis utilizing Pisum sativum peels (Prasad et al., 2017d). Results show that at optimum condition of pH 6, removal of 96% of the dye was attained within one hour with adsorbent dosage of 0.3 g/L. Similarly *Moringa oleifera* leaves were utilized by Prasad et al. (2017c) for synthesizing iron oxide nanoparticles supported on nickel composite for degradation study of Malachite green dye. On analyzing the samples after 25 minutes of reaction through UV-Vis Spectrometer, it was noticed that 90% of the organic dye was removed.

Singh et al. (2017) investigated the removal of methylene blue dye utilizing green tea mediated synthesized iron oxide nanoparticles. The results showed that degradation efficiency of 95% was achieved within 16 min. Ahmed et al. (2014a, 2014b) utilized microwave and microwave-hydrothermal techniques for synthesizing hematite quantum dots, and nanorods respectively. The prepared nanomaterials were utilized for degradation of an organic dye and further the results were compared with commercially available hematite. Average particle size of ionic-liquid synthesized quantum dots, and nanorods were found to be ~10 nm, and 60 nm respectively. Microwave-assisted quantum dots, due to their higher surface area, were found to have better catalytic efficiency as compared to synthesized nanorods, and commercially available hematite.

7.2 HEAVY METAL REMOVAL FROM WASTEWATER

Rao et al. (2013) reported removal of Chromium (VI) utilizing bio Fe^0/Fe_3O_4 nanocomposite and yeast cells. Such prepared nanocomposite proved to be good biosorbents for Cr(VI) removal. The sorption capacity of the iron oxide modified yeast cells was enhanced by three times when compared to the unmodified yeast cells. Better adsorption capacity of modified adsorbent ~186.32 mg/g was reported for 1000 mg/L of Cr(VI) solution under optimum conditions when compared to unmodified adsorbent with adsorption capacity 137.31 mg/g. Moreover, many research studies related to heavy metal removal application through various green synthesized iron-based nanomaterials are reported in Table 7.2.

TABLE 7.2
Green Synthesized Iron-based Nanoparticles for Heavy Metal Removal Applications

Source	Nanoparticle	Morphology	Metal Removed	Efficiency
Syzygium jambos L. Alston leaf extract	nZVI	Spherical, crystalline	Chromium(VI)	99.45% at 90 min
Coconut husk extract	Fe_3O_4 NPs	Amorphous	Calcium and cadmium	Ca ~55%, Cd ~40% at 120 min
Citrus maxima	Fe NPs	Irregular	Chromium(VI)	99.29% at 90 min
Plant extracts: Rosa damascene, *Thymus vulgaris*, and *Urtica dioica*	Fe NPs	Irregular and non-uniform shapes	Chromium(VI)	100% efficiency for all extract mediated NPs at 25 min
Eucalyptus leaf extract	nZVI and Fe-oxide NPs	Spherical	Chromium(VI)	98.9% at 35 min
Tangerine peel extract	Fe-oxide NPs	Spherical	Cadmium	90% at 90 min
Plant leaves	Graphene oxide/Fe-oxide NPs	Spherical NPs assembled over RGO sheets	Lead	96% at 80 min
Eichhornia crassipes leaves	nZVI	Mostly spherical	Chromium(VI)	89.9% at 90 min
Moringa oleifera leaves	Fe-oxide NPs	Nanosphere agglomerate	Fluoride	1.4 mg/g at 40 min
Lantana camara fruit extract	Fe-oxide NPs	Spherical	Nickel	99% at 100 min
Fatty acids in Olive oil	Fe-oxide NPs	Spherical	Nickel	72% at 20 min
Vaccinium corymbosum leaf and shoot extract	Fe (oxide, hydroxide), and nZVI	Spherical, agglomerated	Arsenate	76% at 120 min
Eucalyptus leaf extract	nZVI, Fe-oxide NPs	Spherical,	Chromium(VI) and copper ions	58.9% Cr (VI) and 33.0% Cu (II) at 60 min
Moringa oleifera seed and leaf extract	Fe NPs	Spherical, thick surface layer	Nitrate	For surface water: 70% (leaf extract) 74% (seed extract) at 1440 min
Commercially available tea	Fe/Clay supported	Crystalline	Arsenic (III)	99% at 30 min
Mentha spicata L.	Fe/Chitosan supported	Cubic, dispersed	Arsenic (III), (V)	100% at 30 min

(*Continued*)

TABLE 7.2 (Continued)

Source	Nanoparticle	Morphology	Metal Removed	Efficiency
Eucalyptus globulus	Fe/Chitosan supported	Nono-sized clusters	Arsenic(V)	147 μg/g maximum adsorption capacity
Emblica officinalis	Fe_3O_4-NPs	Uniform spherical	Pb(II) removal	~46.18 mg/g max adsorption capacity
Punica granatum	Fe NPs	Cubic inverse spinel structure	Pb(II) removal	>99.9%
Eucalyptus globulus	nZVI	Spherical	Adsorption of chromium(VI)	98.1% at 30 min
S. jambos L) Oolong tea, A. moluccana	Fe NPs	Spherical	Chromium(VI)	100% at 60 min

Source: [Reproduced with permission from Mondal et al. (2020) © Elsevier].

Synthesis of zero-valent iron nanoparticles (nZVI) using *Euclaptus globules* leaf extract was reported by Madhavi et al. (2013). Utilizing more leaf extract enhanced the reaction rate of nanoparticle formation. Presence of various functional groups along with nature of surface complex was analyzed through FTIR spectroscopy. The synthesized nZVI were applied for adsorption of Cr(VI) metal. At optimum adsorbent dosage of 0.8 g/L highest adsorption efficiency of 98.1% was obtained after 30 minutes of reaction. Furthermore, stable and efficient green synthesized nZVI were reported by Savasari et al. (2015) obtained utilizing ascorbic acid. The prepared nZVI was employed for reduction of Cd(II) by adsorption studies.

Mystrioti et al. (2014, 2015) reported in two different studies, the preparation of nZVI in colloidal suspension utilizing green tea extract with polyphenol coating. The efficiency of synthesized nZVI for chromium removal was studied in groundwater. The average particle diameter of the green synthesized nZVI was found to be 5–10 nm and was effective in removing Cr(VI) from groundwater passing through permeable soil bed. Polyphenolic coating acted as both reducing and capping agent for the prepared nZVI. In order to analyze the effect of contact time on reduction of Cr(VI), different flow rates were maintained in column test study. The study shows, by increasing the contact time the reducing efficiency of Cr(VI) enhances. Chromium is generally insoluble when found in precipitated form as confirmed by leaching tests. Moreover, Cr(VI) in the precipitated form found in test soil was in the range of 280–890 mg/kg of soil, whereas less than 1.4 mg/kg of the soluble Cr was obtained in the soil. Since stable nanoparticles are known to be useful for groundwater remediation purpose, hence nZVI serves this purpose quite well (Mystrioti et al., 2014).

Ponder et al. (2000) reported that by utilizing green synthesized iron nanoparticles heavy metals get removed via redox reaction, and surface adsorption process. The main factor which influences the reactivity of iron nanoparticles depends on the variable oxidation states of iron. Tang and Lo (2013) reported that with the varying oxidation state of any metal the chemical characteristics along with mechanism of reaction between metal and contaminants will vary.

Xiao et al. (2016) in a recent study synthesized iron nanoparticles from various plant leaf extracts and utilized them for chromium removal. The selected plants showed good antioxidant property due to reduction potential activity. The study uses *S. jambos* (L.) alston (SJA) extract, oolong tea (OT) extract, and *A. moluccana* (L.) Willd (AMW) extract. The antioxidant property of the leaf extracts mentioned are in decreasing order respectively. Results showed that 1mL of SJA-Fe NPs solution due to its highest antioxidant property removed Cr(VI) to 91.9% in the first five minutes, whereas complete removal was attained after 60 minutes of reaction. Average particle diameter of the SJA-Fe NPs was ~5 nm and was found to be amorphous in nature.

A similar study involved *Syzygium jambos (L.) Alston* leaf extract which was utilized by Xiao et al. (2017) for synthesizing nZVI to remove hexavalent chromium ions from aqueous solution. Factors such as nanoparticle dosage, solution pH, temperature, and Cr(VI) initial concentration influences the removal efficiency of chromium. Availability of active sites increases with increasing nanoparticles dosage for a fixed concentration of chromium, which in turn enhances the removal rate. Results convey that removal efficiency got enhanced by 15% with temperature increasing from 20°C to 60°C. Due to electrostatic attraction between protonated nanoparticle surface and hexavalent chromium at acidic pH, hence Chromium (VI) removal was found favorable at low pH. The experimental results conveyed that at optimum pH of 5.5, with 0.5 mL of synthesized nanoparticles, and initial chromium concentration of 50 mg/L complete removal of chromium was obtained.

Similarly, Sebastian et al. (2018) developed magnetite nanoparticle using coconut husk extract, and applied for the adsorption of calcium and cadmium present in low levels. Results confirmed that cadmium and calcium removal were more than 40% and 50% respectively at the end of 120 minutes. Experiments suggest that maximum removal was achieved by 30 minutes and reached equilibrium after that. The initial concentration of metal ions was fixed at 50 mg/L, and it was noticed that with increasing adsorbent dosage adsorption rate enhanced until attaining saturation. Since, adsorption of metal ions was involved, hence with increase in solution pH from 2 to 6 the adsorption process enhanced due to electrostatic attraction. Lower and ambient temperatures were found to favor the adsorption process, since at higher temperatures the electrostatic balance between adsorbate metal ions and adsorbent nanoparticle gets distorted.

Tangerine peel extract was utilized by Ehrampoush et al. (2015) for preparing iron oxide nanoparticles, and utilized for removing trace amounts of cadmium present in water bodies. Experimental studies revealed maximum removal of 90% was achieved at pH 4 within 90 minutes. Similarly, in order to remove Pb(II) from water, graphene oxide-based iron oxide nanocomposite was synthesized by Prasad et al. (2017b). Experimental results report to remove Pb(II) about 96% at pH 5, where rapid adsorption occurred within 40 minutes and equilibrium was reached at 80 minutes. The adsorption study was found to be pH dependent, and moreover the adsorbents were reusable by treating with definite concentration acid to desorb Pb(II) ions.

Wei et al. (2017) reported synthesis of iron nanoparticles by using water hyacinth leaves, and utilized it for removal of hexavalent chromium from solution. Maximum removal of 90% Cr(VI) ions was attained within 80 minutes with an initial stock solution maintained at 100 mg/L concentration. The ratio of plant extract and Fe^{3+} salt concentration was maintained at 1:1 for efficient synthesis of iron nanoparticles.

Studies have shown that by utilizing *Lantana camara* fruit extract superparamagnetic iron oxide nanoparticles were synthesized for the removal of Ni(II) ions from aqueous solution (Nithya et al., 2017). Experimental batch studies were performed to optimize the influencing parameters on adsorption process such as stock solution concentration, adsorbent dosage, pH, temperature, and contact time. Due to electrostatic attraction phenomena between adsorbate and adsorbent at high pH, the adsorption process enhanced. An optimum pH 6 was chosen for the process since at high basicity nickel tends to form hydroxides. Higher pH involves the presence of negative ions such as hydroxides and carboxyl with indulges the cation nickel better for adsorption process. Similarly, at lower pH the desorption occurs distorting the electrostatic balance between adsorbate and adsorbent. Sorption capacity enhances with increasing nickel ion concentration in the solution, but on other hand the adsorption efficiency decreases due to lack of active sites. Due to greater surface area possessed by smaller sized adsorbents, adsorption efficiency increases with smaller particle size of nanoparticles.

In a report synthesis of magnetic iron oxide coated with amino-silica support was performed by Mahmoud et al. (2016) utilizing a modified coprecipitation method. Adsorption of heavy metals ions like Pb^{2+}, Cu^{2+}, Cd^{2+} and Hg^{2+} was done by utilizing the synthesized nanocomposite using microwave heating. Microwave heating was performed while keeping the dissolved metal ion, and nanocomposite in contact for a short time. Sorption process of the metal ion occurs on the nanocomposite. Within 15 seconds among the other ions, Pb^{2+} showed better sorption capacity. Experimental results suggest that with increase in temperature desorption of metals did not occur.

7.3 HAZARDOUS POLLUTANT REMOVAL FROM WASTEWATER

Transportation of iron nanoparticles synthesized from green tea extract was investigated by Chrysochoou et al. (2012) by utilizing media such as refined silica sand, and aluminum hydroxide coated sand. Rapid decline in effluent pH was noticed from 8.5 to 2 on injecting green synthesized nZVI. Such a phenomenon was denoted due to the presence of Fe^{3+} as residuary discharge in the solution, along with hydrolysis reactions. Redox potential was observed to increase from

150 mV to 550 mV, instead of the presence of Fe^0 in the synthesized nanoparticles. Such behavior was denoted as a result of polyphenol oxidation available in green tea, which indicates green synthesized nZVI transportation through the medium. There are numerous studies related to wastewater treatment and pollutant removal from wastewater by utilizing various green synthesized iron and iron-based nanomaterials shown in Table 7.3. On a similar note, He and Zhao (2005) investigated the degradation of trichloroethylene (TCE) utilizing bimetallic Fe/Pd nanoparticles synthesized by starch. Less agglomerated starch coated nanoparticles were synthesized which proved to be effective in TCE degradation as compared to those synthesized without stabilizer. A maximum degradation of 98% was obtained within one hour when the reaction was carried out with a nanoparticle dosage of 0.1 g/L.

TABLE 7.3
Green Synthesized Iron-based Nanoparticles for Wastewater Treatment Applications

Source	Nanoparticle	Morphology	Pollutant Removed	Efficiency
Silybum marianum L. (plant seeds)	Copper supports on magnetite nanocatalyst	8.5–60 nm, spherical	Nitrobenzene	95% at 90 min
Euphorbia cochinchensis leaf extract	Iron oxide nanoparticles	100 nm, spherical	2,4-dichloro-phenol	64% at 120 min
Euphorbia bungei boiss leaves	Cu/RGO/Fe_3O_4	RGO sheet dispersed with Fe_3O_4 and Cu NPs	Cyanation of aldehydes	93% at 120 min
Withania coagulans leaf extract	Pd/RGO/Fe_3O_4	<15 nm, spherical	Reduction of 4-nitrophenol	100% at 1 min
Zanthoxylum rhetsa fruit extract	Fe_2O_3@SiO_2	5–21 nm, cluster	Ipso-hydroxy-lation of boronic acid in water	98% at 120 min
Cynometra ramiflora leaf extract	Iron oxide nanoparticles	<100 nm, spherical aggregates	Degradation of Rhodamine B	100% at 15 min
Microwave-assisted synthesis	Silica-supported (MCM-41) iron oxide NPs	<50 nm, spherical	Alkylation of toluene with benzyl chloride	99% at 3 min
Yeast	Protein capped nZVI	3–10 nm, irregular	Pesticide (Dichlorvos)	100% at 60 min
Oak leaves	nZVI	20–100 nm, irregular	Amoxicillin	100% at 95 min

(Continued)

TABLE 7.3 (Continued)

Source	Nanoparticle	Morphology	Pollutant Removed	Efficiency
Grape marc, black tea leaves, vine leaves	nZVI	15–45 nm, spherical agglomerates	Ibuprofen	50%, 50%, 55% at 193 h
Eucalyptus leaves	Fe NPs	20–70 nm, spherical	Ametryn	100% at 135 min
Black tea extract and ionic liquid (N-methyl-butylimidazolium-bromide)	Iron oxide NPs	5–50 nm, irregular	Propranolol	>90% at 40 min
Eucalyptus leaves	Fe NPs	20–80 nm, spheroidal	COD, Total nitrogen and phosphorus	71.7% for total N, 30.4% for total P, and 84.5% for COD at 21 days
Sapindus mukorossi	Iron hexacyano-ferrate	10–60 nm, hexagonal nanorods, spheres and rhombus	PAHs in water and soil (benzo (a) pyrene, chrysene, fluorene, phenanthrene anthracene)	Anthracene, phenan-threne >80%, benzo (a) pyrene, chrysene, fluorine ~70–80% at 48h
Eucalyptus leaf extract	Iron oxide nanoparticles	80–90 nm, spherical	Phosphate	>95% at 100 min
Mangifera indica L. leaf extract	Iron oxide	Nanorods, L: 15 ± 2 nm D: 3.0 ± 0.2 nm	Heavy oil	50% reduction in viscosity
Eichhornia crassipes, Lantana camara and Mimosa pudica leaves	Iron nanoparticles	20–60 nm spherical, 25–50 nm spherical, 65 nm chain- like	Nitrate and phosphate from wastewater	Maximum 74.52% of nitrate and 55.39% of phosphate at 24 h
Starch	Bimetallic Fe/ Pd NPs	14.1–60 nm, Dendritic structure	Trichloro-ethylene	98% at 1 h

Source: [Reproduced with permission from Mondal et al. (2020) © Elsevier].

Wang et al. (2014b) reported the removal of nitrate from wastewater by iron oxide nanoparticles synthesized utilizing green tea and eucalyptus leaves separately. Both syntheses formed spheroidal particles. The study reports a comparison of nitrate removal from iron nanoparticles synthesized by plant materials and chemical synthesis. Results show about 59.7% and 41.4% nitrate removal was obtained utilizing green tea and eucalyptus mediated NPs respectively, whereas nZVI and Fe_3O_4 nanoparticles provided 87.6% and 11.7% removal respectively.

Green synthesized nanoparticles were more stable and preferred over nZVI despite its higher removal efficiency. Experiments were also conducted to compare the reactivity efficiency of the green synthesized Fe NPs and nZVI after two months of exposure. Reactivity efficiency of the green synthesized NPs was found to be retained at 51.7% (green tea) and 40.7% (eucalyptus), whereas the efficacy was decreased by nearly 2.1 fold to 45.4% for nZVI prepared by chemical synthesis.

Groiss et al. (2017) reported the synthesis of iron oxide nanoparticles utilizing *Cynometra ramiflora* leaf extract, and employed it for degradation of Rhodamine B (carcinogen) by Fenton-like catalytic process in the presence of hydrogen peroxide. The degradation was found to be optimum within 15 minutes, when 1.11 mM nanoparticles was utilized with 2% H_2O_2. Such process was found to produce minimum sludge and is thus advantageous to other Fenton-catalyzed processes.

Withania coagulans was reported by Atarod et al. (2016) for synthesizing reduced graphene oxide/iron oxide-based palladium nanoparticles (Pd/RGO/Fe_3O_4). On reacting graphene oxide suspension with *Withania coagulans* leaf extract, along with ferric chloride solution, reduced graphene oxide based iron-oxide nanocomposite was synthesized. In order to separate the RGO/Fe_3O_4 nanocomposite, the mixture was stirred, refluxed and magnetically separated. Finally, in order to obtain Pd/RGO/Fe_3O_4 nanocomposite with an average particle size of 7–13 nm, palladium chloride was very carefully introduced into the as-prepared nanocomposite. The catalytic degradation effect of the prepared nanocomposite toward 4-nitrophenol in the presence of sodium borohydride was studied. Fading of the yellow color was obtained during the reaction in the presence of nanocatalyst while forming 4-aminophenol. Whereas no such appreciable reduction was observed in the absence of the nanocatalyst. The study further conveys that Pd/RGO/Fe_3O_4 nanocomposite had a better catalytic property compared to RGO/Fe_3O_4 nanocomposite, which indicates the enhancing property of palladium nanoparticles.

Wang et al. (2014b) reported the removal study of total nitrogen, phosphorus and COD from poultry breeding wastewater utilizing iron nanoparticles synthesized from eucalyptus leaf extract. The study shows that a maximum removal of 71.7%, 30.4% and 84.5% was obtained for total nitrogen, phosphorus and COD respectively. Similarly, iron nanoparticles were synthesized by Wei et al. (2016) utilizing *Citrus maxima* peel extract for the removal of Cr(VI). Optimum *Citrus maxima* peel extract: Iron (III) solution of 1:3 was maintained to synthesize better property iron nanoparticles which resulted in complete Cr(VI) removal within 90 minutes. Similarly, removal of Cr(VI) was also reported by Fazlzadeh et al. (2017) using iron nanoparticles synthesized from three different plant extracts *Rosa damascene*, (RD), *Thymus vulgaris* (TV), and *Urtica dioica* (UD). Experimental studies show greater than 90% efficiency was obtained within ten minutes. Shanker et al. (2017) reported the degradation of eight selected priority polycyclic aromatic hydrocarbons (PAHs) in soil and water. Initial such high molecular weight PAHs varies from 50 to 250 mg $L-^{1}$. The optimum experimental condition at pH of 7.0 resulted in 80–90% degradation of anthracene and phenanthrene within 48 hours.

Mehrotra et al. (2017) employed protein capped zero-valent iron NPs synthesized using yeast, for complete degradation of organophosphorus insecticide (Dichlorvos). At an optimum experimental condition with 2000 mg/L nanoparticles dosage, 1000 μL H_2O_2 a maximum degradation of 99.9% was obtained within 60 minutes of reaction.

Moreover, copper-supported iron oxide nanoctalyst was synthesized by Sajadi et al. (2016) for studying the reduction of nitroarenes in wastewater. The degradation process was employed with sodium borohydride acting as reducing agent. Experimental studies conveyed the fact that the prepared nanocatalyst along with $NaBH_4$ obtained a maximum degradation of 90% within 90 minutes, whereas no degradation was observed in presence of $NaBH_4$ alone.

Machado et al. (2017) reported the synthesis of nanoscale zerovalent iron utilizing oak leaf extract for studying the degradation of the common antibiotic, amoxicillin, in wastewater. With an optimum nanoparticle to analyte ratio of 15:1, about 100% degradation was observed in aqueous solution within 95 minutes of reaction time. Whereas, about 1800 hours was required at the same condition for 53% degradation to occur in the soil, due to nanoparticle soil matrix interaction.

Al-Ruqeishi et al. (2016) reported the treatment of high viscous oil utilizing iron oxide nanorods synthesized using mango tree leaf extract. Experimental study involved the blending of iron oxide nanoparticles with heavy crude oil, and microwaving for 80 seconds. An appreciable reduction in viscosity up to 50% was obtained with 0.6 g/L nanorods at 30–50°C.

7.4 GREEN SYNTHESIZED IRON NPs IMMOBILIZED ON POLYMER AND OTHER SUPPORTS FOR POLLUTANT DEGRADATION

Since nanoparticles are prone to agglomeration which reduces its effectiveness towards catalytic and adsorptive activity, such limitation was overcome by immobilizing nanoparticles within polymeric, zeolitic, and silica-based supports. Smuleac et al. (2011) utilized polyvinylidene fluoride (PVDF) membrane as a support to incorporate Fe and bimetallic Fe/Pd nanoparticles through the green synthesis technique. Through polymerization reaction, PVDF membranes were modified with polyacrylic acid (PAA), whereas for synthesizing Fe and Fe/Pd nanoparticles, green tea extract served as the reducing agent. Nanoparticle embedded PVDF/PAA membrane was found to incorporate nanoparticles with an average particle diameter of 20–30 nm. The synthesized nanoparticle impregnated membrane was utilized for studying the degradation of toxic organic pollutant, trichloroethylene (TCE). Catalytic reactivity of the membrane dechlorinated TCE compound, and was found to increase with more nanoparticle incorporation within the membrane. Moreover, it was reported that bimetallic Fe/Pd incorporation enhanced the catalytic process, rather than only Fe nanoparticles.

In another study, Tandon et al. (2013) synthesized zero-valent iron (nZVI) nanoparticles using tea liquor and stabilized on montmorillonite K10. The synthesized nZVI had an average particle size of 59.08 ± 7.81 nm. The nZVI impregnated montmorillonite support was then employed for an arsenic removal study. Experimental observations revealed 90% removal of As(III) within 30 minutes of reaction at both high and low pH of 2.75 and 11.1 respectively. It was observed that nZVI impregnated montmorillonite K10 resulted for better As(III) removal than only montmorillonite K10.

Prasad et al. (2014) reported the synthesis of iron nanoparticles by *Mentha spicata L.* leaf extract, supported on chitosan composite, and studied its efficiency in removal of arsenite(III) and arsenate(V) from aqueous solution. The average particle diameter obtained through the transmission electron microscope (TEM) ranged from 20 to 45 nm. The nanoparticle resembled a core-shell structure with functional groups like N–H, C=O, C=C and C=N present on the surface. The results conveyed that 2 g/L nanoparticle-chitosan composite was efficient in removing 100 mg/L As(III) and As(V) at about 98.79% and 99.65%, respectively at the end of 60 minutes.

Martínez-Cabanas et al. (2016) studied different extracts like chestnut tree (*Castanea sativa*), eucalyptus (*Eucalyptus globulus*), gorse (*Ulex europaeus*), and pine (*Pinus pinaster*), for synthesizing iron nanoparticles. The study reported the effect of plant extract ratio on the formation of iron nanoparticles, and the high antioxidant property present in eucalyptus was found to be most suitable for nanoparticle synthesis. Prepared nanoparticles were mixed with chitosan, and furthermore the chitosan beads were characterized. Iron nanoparticle encapsulated with chitosan was utilized for As(V) removal. Moreover, maximum adsorption capacity of 147 µg/g was achieved with 25 g/L adsorbent dosage at pH 6–7.

Saikia et al. (2017) reported synthesis of iron oxide nanoparticles supported on silica through a cheap green method, and utilized it further as a catalyst for preparing phenol from boronic acids through ipso-hydroxylation reaction, instead of using hydrogen peroxide (H_2O_2). Experimental studies show about 98% yield in aqueous solution when about 4 mg of nanocatalyst was utilized within two hours at 50°C.

Zheng et al. (2014), in order to extract organophosphorus pesticides from solution, prepared nanosorbent of iron oxide and ionic liquid upon a polymeric support. Methanol proved to be the better dispersion medium for the prepared nanosorbents, as well as effective elution solvent. At an optimum condition of 2.4 g/L adsorbent dosage maximum yield of 81.4–112.6% was obtained at the end of 80 minutes.

Mondal and Purkait (2017) synthesized iron nanoparticles with an average diameter of 32 nm by employing cardamom extract. Furthermore, the nanoparticles were incorporated within pH responsive poly(vinylidene fluoride-co-hexafluoro propylene) (PVDF-co-HFP) by a mixing method. The prepared nanoparticle embedded membrane was studied for nitrobenzene (NB) reduction purpose by varying iron NP content (wt%), pH, and time (min). At an optimum pH 11.9, iron content 0.01 wt%, and NB degradation of 57% was obtained at the end of 50 minutes. Aniline conversion of 9.65 ppm was obtained from 100 ppm nitrobenzene.

In another work, Mondal and Purkait (2018) utilized iron oxide incorporated pH responsive polymeric membrane for degradation of nitrobenzene and fluoride rejection purpose simultaneously. In this work, clove extract was employed for synthesizing iron NPs with an average diameter of 13.5 nm having core shell structure with Fe^0 at core (Figure 7.1). At optimized condition of pH = 3, iron content 0.01 wt%, and time 39 min, NB reduction of 86.7% was obtained with aniline formation of 12.7 ppm. Fluoride rejection was highest for 20 ppm stock solution of 72% at optimized conditions.

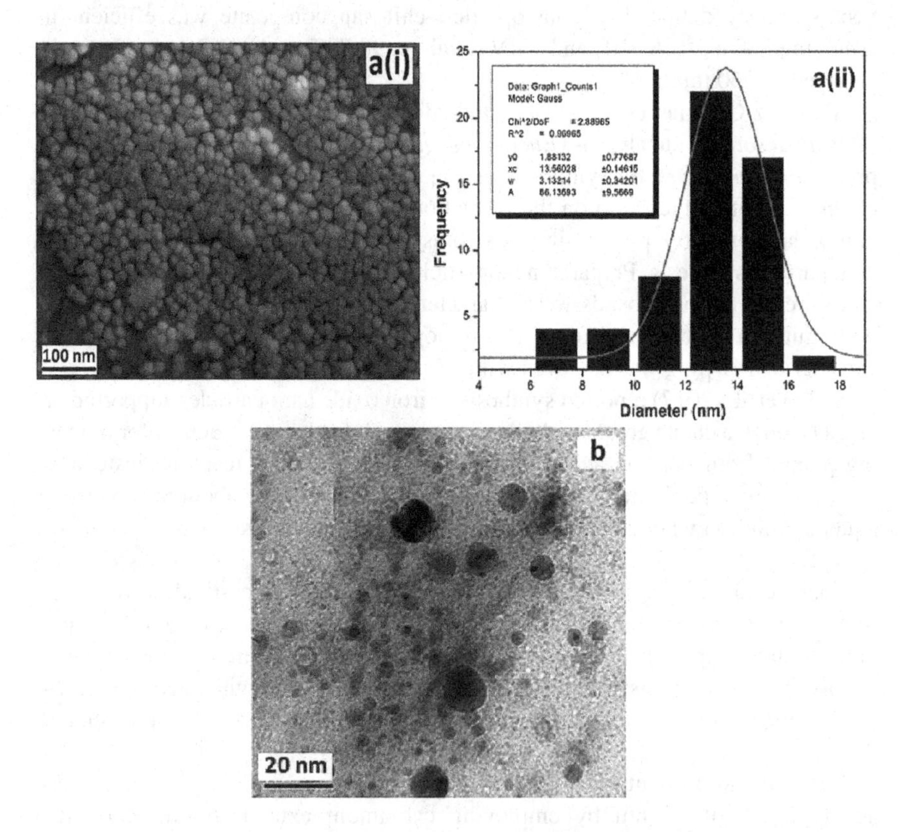

FIGURE 7.1 (a) FESEM analysis and (b) TEM analysis of prepared iron NPs [Reproduced with permission from Mondal and Purkait (2018) © Elsevier].

REFERENCES

Ahmed, F., Arshi, N., Anwar, M.S., Danish, R., Koo, B.H., 2014a. Quantum-confinement induced enhancement in photocatalytic properties of iron oxide nanoparticles prepared by ionic liquid. *Ceram. Int.* 40 (10), 15743–15751.

Ahmed, T., Imdad, S., Yaldram, K., Butt, N.M., Pervez, A., 2014b. Emerging nanotechnology-based methods for water purification: a review. *Desalin. Water Treat.* 52 (22–24), 4089–4101.

Al-Ruqeishi, M.S., Mohiuddin, T., Al-Saadi, L.K., 2016. Green synthesis of iron oxide nanorods from deciduous Omani mango tree leaves for heavy oil viscosity treatment. *Arab. J. Chem.* 12 (8), 4084–4090.

Atarod, M., Nasrollahzadeh, M., Mohammad Sajadi, S. 2016. Green synthesis of Pd/RGO/Fe_3O_4 nanocomposite using *Withania coagulans* leaf extract and its application as magnetically separable and reusable catalyst for the reduction of 4-nitrophenol. *J. Colloid Interface Sci.* 465, 249–258.

Chrysochoou, M., McGuirea, M., Dahalb, G. 2012. Transport characteristics of green-tea nano scale zero valent iron as a function of soil mineralogy. *Chem. Eng. Trans.* 28, 122–126.

Ehrampoush, M.H., Miria, M., Salmani, M.H., Mahvi, A.H. 2015. Cadmium removal from aqueous solution by green synthesis iron oxide nanoparticles with tangerine peel extract. *J. Environ. Health Sci. Eng.* 13 (1), 84–90.

Fazlzadeh, M., Rahmani, K., Zarei, A., Abdoallahzadeh, H., Nasiri, F., Khosravi, R. 2017. A novel green synthesis of zero valent iron nanoparticles (NZVI) using three plant extracts and their efficient application for removal of Cr(VI) from aqueous solutions. *Adv. Powder Technol.* 28 (1), 122–130.

Groiss, S., Selvaraj, R., Varadavenkatesan, T., Vinayagam, R. 2017. Structural characterization, antibacterial and catalytic effect of iron oxide nanoparticles synthesised using the leaf extract of *Cynometra ramiflora. J. Mol. Struct.* 1128, 572–578.

He, F., Zhao, D. 2005. Preparation and characterization of a new class of starch-stabilized bimetallic nanoparticles for degradation of chlorinated hydrocarbons in water. *Environ. Sci. Technol.* 39, 3314–3320.

Hoag, G.E., Collins, J.B., Holcomb, J.L., Hoag, J.R., Nadagouda, M.N., Varma, R.S. 2009 Degradation of bromothymol blue by 'greener' nano-scale zero-valent iron synthesized using tea polyphenols. *J. Mater. Chem.* 19, 8671–8677.

Huang, L., Luo, F., Chen, Z., Megharaj, M., Naidu, R. 2015. Green synthesized conditions impacting on the reactivity of Fe NPs for the degradation of malachite green. *Spectrochim. Acta A* 137, 154–159.

Huang, L., Weng, X., Chen, Z., Megharaj, M., Naidu, R. 2013. Synthesis of iron-based nanoparticles using oolong tea extract for the degradation of malachite green. *Spectrochim. Acta A* 117, 801–804.

Kuang, Y., Wang, Q., Chen, Z., Megharaj, M., Naidu, R. 2013. Heterogeneous Fenton-like oxidation of monochlorobenzene using green synthesis of iron nanoparticles. *J. Colloid Interface Sci.* 410, 67–73.

Luo, F., Chen, Z., Megharaj, M., Naidu, R. 2014 Biomolecules in grape leaf extract involved in one-step synthesis of iron-based nanoparticles. *RSC Adv.* 4, 53467–53474.

Machado, S., Pacheco, J.G., Nouws, H.P.A., Albergaria, J.T., Delerue-Matos, C. 2017. Green zero-valent iron nanoparticles for the degradation of amoxicillin. *Int. J. Environ. Sci. Technol.* 14 (5), 1109–1118.

Madhavi, V., Prasad, T.N., Reddy, A.V., Ravindra Reddy, B., Madhavi, G. 2013. Application of phytogenic zerovalent iron nanoparticles in the adsorption of hexavalent chromium. *Spectrochim. Acta A* 116, 17–25.

Mahmoud, M.E., Amira, M.F., Zaghloul, A.A., Ibrahim, G.A.A. 2016. Microwave-enforced sorption of heavy metals from aqueous solutions on the surface of magnetic iron oxide-functionalized-3-aminopropyltriethoxysilane. *Chem. Eng. J.* 293, 200–206.

Martínez-Cabanas, M., López-García, M., Barriada, J.L., Herrero, R., Sastre de Vicente, M.E. 2016. Green synthesis of iron oxide nanoparticles. Development of magnetic hybrid materials for efficient As(V) removal. *Chem. Eng. J.* 301, 83–91.

Mehrotra, N., Tripathi, R.M., Zafar, F., Singh, M.P. 2017. Catalytic degradation of dichlorvos using biosynthesized zero valent Iron nanoparticles. *IEEE Trans. Nanobiosci.* 16 (4), 280–286.

Mondal, P., Purkait, M.K. 2017. Green synthesized iron nanoparticle-embedded pH-responsive PVDF-co-HFP membranes: Optimization study for NPs preparation and nitrobenzene reduction. *Sep. Sci. Technol.* 52 (14), 2338–2355.

Mondal, P., Purkait, M.K. 2018. Green synthesized iron nanoparticles supported on pH responsive polymeric membrane for nitrobenzene reduction and fluoride rejection study: Optimization approach. *J. Clean. Prod.* 170, 1111–1123.

Mondal, P., Anweshan, A., Purkait, M.K. 2020. Green synthesis and environmental application of iron-based nanomaterials and nanocomposite: A review. *Chemosphere.* 259, 127509.

Mystrioti, C., Papassiopi, N., Xenidis, A., Dermatas, D., Chrysochoou, M. 2015. Column study for the evaluation of the transport properties of polyphenol-coated nanoiron. *J. Hazard. Mat.* 281, 64–69.

Mystrioti, C., Xenidis, A., Papassiopi, N. 2014. Reduction of hexavalent chromium with polyphenol-coated nano zero-valent iron: Column studies. *Desal. Water Treat.* 56, 1162–1170.

Nithya, K., Sathish, A., Senthil Kumar, P., Ramachandran, T. 2017. Fast kinetics and high adsorption capacity of green extract capped superparamagnetic iron oxide nanoparticles for the adsorption of Ni(II) ions. *J. Ind. Eng. Chem.* 59, 230–241.

Njagi, E.C., Huang, H., Stafford, L., Genuino, H., Galindo, H.M., Collins, J.B., Hoag, G.E., Suib, S.L. 2011. Biosynthesis of iron and silver nanoparticles at room temperature using aqueous sorghum bran extracts. *Langmuir* 27, 264–271.

Ponder, S.M., Darab, J.G., Mallouk, T.E. 2000. Remediation of Cr (VI) and Pb (II) aqueous solutions using supported, nanoscale zero-valent iron. *Environ. Sci. Technol.* 34, 2564–2569.

Prasad, C., Karlapudi, S., Venkateswarlu, P., Bahadur, I., Kumar, S. 2017a. Green arbitrated synthesis of Fe_3O_4 magnetic nanoparticles with nanorod structure from pomegranate leaves and congo red dye degradation studies for water treatment. *J. Mol. Liq.* 240, 322–328.

Prasad, C., Krishna Murthy, P., Hari Krishna, R.H., Sreenivasa Rao, R., Suneetha, V., Venkateswarlu, P. 2017b. Bio-inspired green synthesis of RGO/Fe_3O_4 magnetic nanoparticles using *murraya koenigii* leaves extract and its application for removal of Pb (II) from aqueous solution. *J. Environ. Chem. Eng.* 5 (5), 4374–4380.

Prasad, C., Sreenivasulu, K., Gangadhara, S., Venkateswarlu, P. 2017c. Bio-inspired green synthesis of Ni/Fe_3O_4 magnetic nanoparticles using *Moringa oleifera* leaves extract: A magnetically recoverable catalyst for organic dye degradation in aqueous solution. *J. Alloys. Compd.* 700, 252–258.

Prasad, C., Yuvaraja, G., Venkateswarlu, P. 2017d. Biogenic synthesis of Fe_3O_4 magnetic nanoparticles using *Pisum sativum* peels extract and its effect on magnetic and methyl orange dye degradation studies. *J. Magn. Magn. Mater.* 424, 376–381.

Prasad, K.S., Gandhi, P., Selvaraj, K. 2014. Synthesis of green nano iron particles (GnIP) and their application in adsorptive removal of As(III) and As(V) from aqueous solution. *Appl. Surf. Sci.* 317, 1052–1059.

Rao, A., Bankar, A., Kumar, A.R., Gosavi, S., Zinjarde, S. 2013. Removal of hexavalent chromium ions by *Yarrowia lipolytica* cells modified with phyto-inspired Fe^0/Fe_3O_4 nanoparticles. *J. Contam. Hydrol.* 146, 63–73.

Saikia, I., Hazarika, M., Hussian, N., Das, M.R., Tamuly, C. 2017. Biogenic synthesis of $Fe_2O_3@SiO_2$ nanoparticles for ipso-hydroxylation of boronic acid in water. *Tetrahedron Lett.* 58 (45), 4255–4259.

Sajadi, S.M., Nasrollahzadeh, M., Maham, M. 2016. Aqueous extract from seeds of *Silybum marianum* L. as a green material for preparation of the Cu/Fe_3O_4 nanoparticles: A magnetically recoverable and reusable catalyst for the reduction of nitroarenes. *J. Colloid Interface Sci.* 469, 93–98.

Savasari, M., Emadi, M., Bahmanyar, M.A., Biparva, P. 2015. Optimization of Cd(II) removal from aqueous solution by ascorbic acid-stabilized zero-valent iron nanoparticles using response surface methodology. *J. Ind. Eng. Chem.* 21, 1403–1409.

Sebastian, A., Nangia, A., Prasad, M.N.V. 2018. A green synthetic route to phenolics fabricated magnetite nanoparticles from coconut husk extract: implications to treat metal contaminated water and heavy metal stress in *Oryza sativa* L. *J. Clean. Prod.* 174, 355–366.

Shahwan, T., Abu Sirriah, S., Nairat, M., Boyacı, E., Erŏglu, A.E., Scott, T.B., Hallam, K.R. 2011. Green synthesis of iron nanoparticles and their application as a Fenton-like catalyst for the degradation of aqueous cationic and anionic dyes. *Chem. Eng. J.* 172, 258–266.

Shanker, U., Jassal, V., Rani, M., 2017. Green synthesis of iron hexacyanoferrate nanoparticles: Potential candidate for the degradation of toxic PAHs. *J. Environ. Chem. Eng.* 5 (4), 4108–4120.

Singh, K.K., Senapati, K.K., Sarma, K.C., 2017. Synthesis of superparamagnetic Fe_3O_4 nanoparticles coated with green tea polyphenols and their use for removal of dye pollutant from aqueous solution. *J. Environ. Chem. Eng.* 5 (3), 2214–2221.

Smuleac, V., Varma, R., Sikdar, S., Bhattacharyya, D. 2011. Green synthesis of Fe and Fe/Pd bimetallic nanoparticles in membranes for reductive degradation of chlorinated organics. *J. Membr. Sci.* 379, 131–137.

Tandon, P.K., Shukla, R.C., Singh, S.B. 2013. Removal of arsenic(III) from water with clay-supported zerovalent iron nanoparticles synthesized with the help of tea liquor. *Ind. Eng. Chem. Res.* 52, 10052–10058.

Tang, S.C., Lo, I.M. 2013. Magnetic nanoparticles: essential factors for sustainable environmental applications. *Water Res.* 47, 2613–2632.

Wang, T., Lin, J., Chen, Z., Megharaj, M., Naidu, R. 2014b. Green synthesized iron nanoparticles by green tea and eucalyptus leaves extracts used for removal of nitrate in aqueous solution. *J. Clean. Prod.* 83, 413–419.

Wang, Z. 2013. Iron complex nanoparticles synthesized by eucalyptus leaves. *ACS Sustain. Chem. Eng.* 1, 1551–1554.

Wang, Z., Fang, C., Megharaj, M. 2014a Characterization of iron-polyphenol nanoparticles synthesized by three plant extracts and their Fenton oxidation of azo dye. *ACS Sustain. Chem. Eng.* 2, 1022–1025.

Wei, Y., Fang, Z., Zheng, L., Pokeung, E., 2017. Biosynthesized iron nanoparticles in aqueous extracts of *Eichhornia crassipes* and its mechanism in the hexavalent chromium removal. *Appl. Surf. Sci.* 399, 322–329.

Wei, Y., Fang, Z., Zheng, L., Tan, L., Tsang, E.P. 2016. Green synthesis of Fe nanoparticles using *Citrus maxima* peels aqueous extracts. *Mater. Lett.* 185, 384–386.

Xiao, Z., Yuan, M., Yang, B., Liu, Z., Huang, J., Sun, D. 2016. Plant-mediated synthesis of highly active iron nanoparticles for Cr(VI) removal: Investigation of the leading biomolecules. *Chemosphere* 150, 357–364.

Xiao, Z., Zhang, H., Xu, Y., Yuan, M., Jing, X., Huang, J., et al., 2017. Ultra-efficient removal of chromium from aqueous medium by biogenic iron based nanoparticles. *Sep. Purif. Technol.* 174, 466–473.

Zheng, X., He, L., Duan, Y., Jiang, X., Xiang, G., Zhao, W., Zhang, S. 2014. Poly(ionicliquid) immobilized magnetic nanoparticles as new adsorbent for extraction and enrichment of organo-phosphorus pesticides from tea drinks. *J. Chromatogr. A.* 1358, 39–45.

8 Biomedical and Diagnostic Applications of Iron-based Nanomaterials

8.1 INTRODUCTION

The exclusive characteristics of iron oxide nanoparticles (IONPs) have demonstrated huge potential for a wide range of biomedical uses. They have shown potential to be useful in applications such as disease detection, imaging, hyperthermia, magnetic separation, cell proliferation in repairing tissues, and even in drug delivery. Even though nanostructures cobalt, nickel, and iron are famous for demonstrating super-paramagnetic properties along with high magnetic susceptibility, IONPs like hematite (α-Fe_2O_3), magnetite (Fe_3O_4), and maghemite (γ-Fe_2O_3) are also widely investigated magnetic nanoparticles. Due to their intrinsic characteristics these IONPs show aggregation nature in the influence of a magnetic field. Additionally, due to the fact that they are more stable in colloid state, they are more biocompatible, and exhibit certain required magnetic properties which might make them superior candidates for several biomedical applications (Huang et al., 2009; Karimzadeh et al., 2017). IONPs may vary from each other due to the availability of both Fe^{3+} and Fe^{2+} ions. Here, the available divalent ions are located at the octahedral sites, whereas the trivalent ions are spread throughout the tetrahedral and octahedral orientations. Further, the α-Fe_2O_3 consists of Fe^{3+} ions spread at the octahedral positions and for the γ-Fe_2O_3 these are available in the octahedral and tetrahedral location, whereas the Fe^{2+} vacant spaces are available at the octahedral sites (Wu et al., 2015). Because of their possible polymorphism and electron hopping ability, these IONPs have been considered useful in biological as well as technical applications.

Additionally, a Janus formation can be obtained using half of the IONPs with the other half as part of the functional material, followed by sandwiching the IONPs between the two functional materials so as to obtain an unique shell–core–shell structure (Wu et al., 2015). Other relevant techniques such as thermal putrefaction, electrochemical, sol-gel, microwave, micro-emulsion, coprecipitation, hydrothermal, sonochemical and biosynthesis were developed to fabricate IONPs (Huber, 2005). Furthermore, the shape and size of the IONPs along with their magnetism also plays a decisive role in deciding their intrinsic

DOI: 10.1201/9781003243632-8

properties. For instance, the Fe_3O_4 and Fe_2O_3 NPs exhibit the various ferrimagnetisms at ambient temperature. The IONPs also loses their diffusivity with time because of the agglomeration of the particles and also lose their magnetic nature because of the oxidation process. Henceforth, various involved techniques are used to optimize the developed NPs in an inanimate atmosphere and efforts are made to make them dissolvable in water at physiological pH, for their use in nanomedicine (Wu et al., 2015). For application in several biomedical fields, including in vivo studies, parameters like biocompatibility, low toxicity, higher retentiveness, biodegradability, and magnetism perform a decisive role in positioning the IONPs at the desired sites. Further, for the detection of various diseases, IONPs play as a probe in techniques like magnetic resonance imaging (MRI), positron emission tomography (PET) and near-infrared fluorescence (NIRF) imaging (Ju et al., 2017; Xie et al., 2010). Contrarily, the IONPs turned out to be attractive for therapeutic nano-medicine from treating cancer to antimicrobial activity (Nehra et al., 2018; Patra et al., 2017). Further, IONPs were used as nanocarriers for theranostics, for improving drug activity in coupling therapy. Henceforth, this book chapter focuses on various IONPs applications in the biomedical field and their role in biomedicine, diagnostics, and therapeutics.

8.2 ANTICANCER DRUG DELIVERY

Figure 8.1 demonstrates a drug delivery system for a specific site, where the Fe_3O_4-NPs loaded drug is ingested by a human via parenteral drug administration. As can be seen, the drug loaded Fe_3O_4-NPs are injected into the human body via

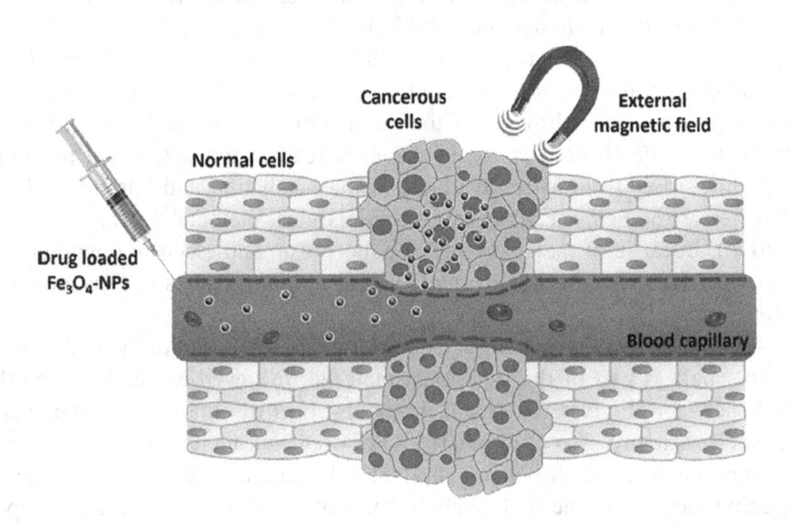

FIGURE 8.1 Targeted drug delivery system using drug loaded Fe_3O_4-NPs. [Reproduced with permission from Yew et al. (2020) © Elsevier].

the blood capillary and are made to arrive at the specified location using external magnetic field. This helps to agglomerate the drug and administer it to the desired site, and hence enhance the cancer curing process without affecting the neighboring healthy cells. Surface modification of Fe_3O_4-NPs was introduced using a wide range of materials including chitosan, polymer, and silica. Studies successfully demonstrate the potential of Fe_3O_4-NPs for this method of anticancer drug delivery.

Further, in vitro drug delivery investigation is even more crucial, as it is important to recognize the functioning of the nanocarriers in the body to eliminate the cancer cells. Several studies have been done on mice as hypodermic tumor models. In a similar study, a pH-sensitive dual targeting magnetic nanocarrier applicable for chemo-phototherapy for curing was prepared by Lu et al. (2018). They procured the magnetic graphene oxide (MGO) by attaching Fe_3O_4-NPs on graphene oxide (GO) with the help of chemical coprecipitation method. Later, the MGO was further doped with polyethylene glycol (PEG) and cetuximab (CET) to obtain MGO-PEG-CET. Further, an anticancer drug called doxorubicin (DOX) was added to MGO-PEG-CET to form MGO-PEG-CET/DOX for anticancer study. Studies were carried out to understand the antitumor efficacy of MGO-PEG-CET/DOX in vivo in the xenograft tumor model in mice. The experiment involved BALB/c with CT-26 tumors of 60–100 mm^3 beneath the skin, treatable using normal saline (control) and DOX in various ways. The observation was done with images of the tumor-bearing mice from day 0 and 14 and the tumor size variation was recorded. On day 14 the tumor was removed from the mice depicting the treatment stages and curing process at a different degree. On the 14th day the tumor tissue witnessed H&E staining and the results demonstrated that necrosis of the cancer cells successful in MGO-PEG-CET/DOX + magnet and MGO-PEG-CET/DOX + laser + group magnet. However, the cells continued to grow for control, DOX and MGO-PEG-CET/DOX groups. To study the reduction in tumor size the images were taken on adaily basis and a graph for relative tumor volume post normalizing the tumor volume was obtained. MGO-PEG-CET/DOX + magnet and MGO-PEG-CET/DOX + magnet + laser demonstrated convincing tumor suppression during the observation period ($*p < 0.05$), in comparison to the control. Further, DOX and MGO-PEG-CET/DOX groups demonstrated reductions in tumor volume, however, no notable variation in the volume of the tumor with respect to the control process throughout the experiment was observed. The study therefore highlights the importance of dual targeting along with external magnetic guidance, however, the MGO-PEG-CET/DOX + magnet treatment could not reduce tumor growth after day eight and a rapid rise in the volume of the tumor was observed. This led to the use of laser light as photo-thermal therapy to reduce tumor growth. MGO-PE G-CET/DOX + magnet + laser therapy may control the tumor's growth, and can also reduce the tumor's size. In the control group, better weight gain in the mice, compared with the group receiving DOX treatment, was observed. The adverse effect from chemotherapy may be attributed as the major reason for this; however, there was negligible change in the appetite and behavior of the mice (Figure 8.2).

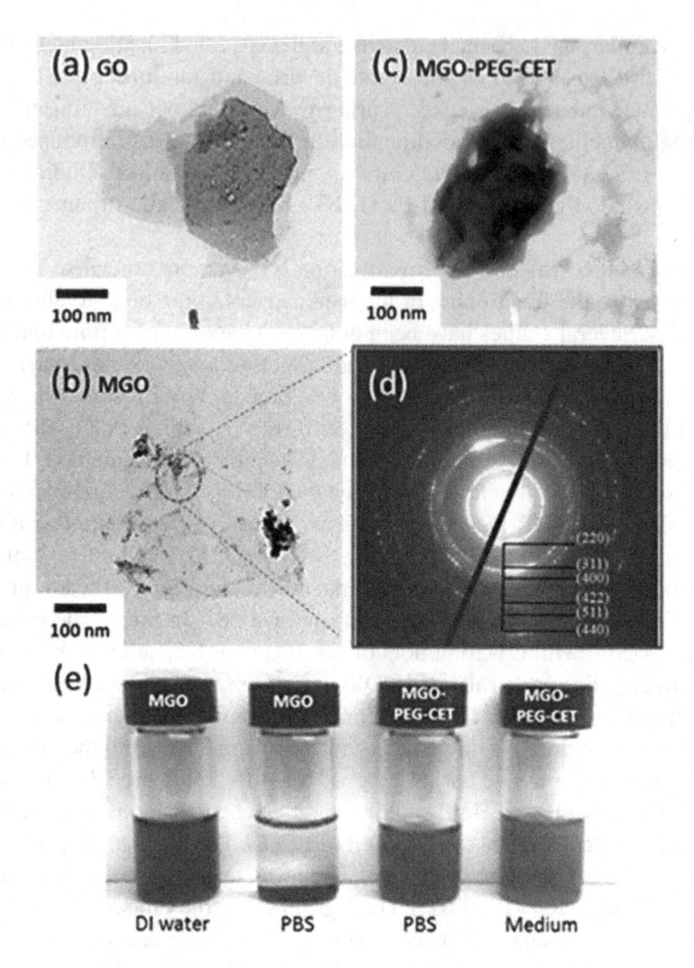

FIGURE 8.2 Transmission electron images of graphene oxide (GO) (a); MGO (b); and, MGO-PEG-CET after staining with 2% phosphotungstic acid (c); (d) The selected area electron diffraction patterns of MGO in the circled area in (b); (e) The suspension stability of 0.1 mg/mL MGO and MGO-PEG-CET in deionized (DI) water, phosphate buffered saline (PBS) and cell culture medium after 24 h [Reproduced with permission from Lu et al. (2018) © MDPI].

8.3 CONTRAST AGENT IN MAGNETIC RESONANCE IMAGING

MRI uses magnetic fields gradients, electrical fields, and utilizes radio waves to explicate the body's detailed internal structure. It finds wide range of uses in diagnosing several diseases and disorders of crucial body organs including the brain, blood vessels, heart, liver, etc. With advancement in the technology, developments allowed the creation of superior magnetic contrast agents like gadolinium, superparamagnetic iron oxides, superparamagnetic iron oxides of extremely small size (5–10 nm), carbon nanotubes modified with gadolinium, paramagnetic micelles incorporated with quantum dots, and several soft NPs like liposomes, ferumoxytol-facilitated. MRI was found to be more susceptible for detecting necrosis in tumor cells in the early stages. Studies also suggest that ferumoxytol attractedrenal failure sufferers as a substitute to gadolinium-based contrast carriers for vascular MRI (Hope et al., 2015). As the feraheme is considered by macrophages in liver, spleen, and lymph nodes they might be further investigated for imaging tumors, macrophages, vascular lesions, and other organs.

IONPs are unique materials, which can be optimized for multifunctional uses. In a similar attempt, Xie et al. (2010) synthesized dopamine doped IONPs, which were further enclosed inside the human serum albumin (HSA) matrices. Studies suggest that HSA coated IONPs can be further labeled with dyes, namely Cy5.5 and ^{64}Cu-DOTA, enabling multiplexed imaging capability and assessed in a subcutaneous U87MG xenograft mouse model. Reports suggest that a triple modality imaging involving MRI, NIRF and PET could be viable within ex vivo and in vivo

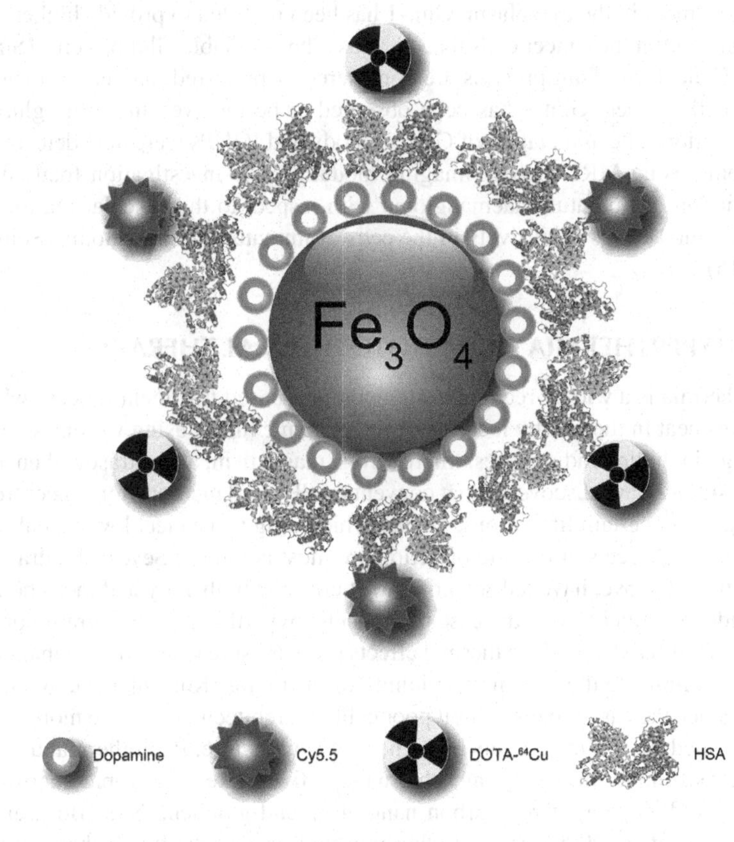

FIGURE 8.3 Schematic illustration of the multifunctional HSA-IONPs. The pyrolysis-derived IONPs were incubated with dopamine, after which the particles became moderately hydrophilic and could be doped into HSA matrices in a way similar to drug loading [Reproduced with permission from Xie et al. (2010) © Elsevier].

investigational states. HSA coating demonstrated extension of the time of blood circulation, better extravasation and agglomeration in desired tissues, and relatively declined uptake in macrophages in the areas near the tumor. Figure 8.3 shows the schematic illustration of the multifunctional HSA-IONPs.

Recent investigations reveal image-guided photothermal therapy (PTT) to have better substitute therapeutic modality than widely used traditional methods. It is further assumed to have capacity to provide an advanced precision therapy alternative. With current technological advancements in material science, materials with a desired composition and of specific shape and size may be fabricated. However, synthesizing multicomponent materials along with the required specifications still presents challenges to overcome. Malignant cells have a significantly better metabolism rate and higher glucose uptake. This mechanism is used to observe the cancer tissues under PET, consuming enhanced radiolabeled glucose analogy uptake, [18F]-2-fluoro-2-deoxy-d-glucose. Glucose transporter (Glut) proteins are available in the mammalian cells plasma membrane, providing glucose movement in the cytoplasm. Glut-1 has been reported to provide higher glucose transport in the cancer cells than various other available Glut proteins (Singh 2017). Henceforth, Glut proteins are considered as preferred markers to identify cancer cells/tissues. Glut-1 has been observed to be involved in higher glucose transportation. The association of Glut antibody and IONPs were able detect haemangioma using MRI contrast imaging modality. The investigation focused on the variation of immature haemangioma with respect to the vascular malformation, as Glut-1 exists exclusively in the cells of immature haemangioma (Sohn et al., 2015).

8.4 HYPERTHERMIA AND PHOTOTHERMAL THERAPY

Hyperthermia is a widely recognised temperature-based treatment process which generates heat in the nearby location or ina systemic tumor using various sources of energy including radio waves, microwaves, magnetism, and ultrasound energy. Recent studies have discovered that conventionally used methods for cancer treatments possess certain limitations, among which drug resistance, low availability, and drug side effects at the site of action are most common. Several hindrances, as mentioned above, have led scientists to couple chemotherapy and radiotherapy alongside hyperthermia. In the case of magnetic hyperthermia, the intratumorally administered IONPs produce thermal effects postexposure to an external magnetic field, and eliminate the cells in the vicinity of the tumor (Kolosnjaj and Wilhelm, 2017). Since the cancer cells exhibit poor cellular architecture they are more likely to be harmed by less increments in the nearby temperature. Photothermal therapy may be feasible by impregnating other nanoparticles like anisotropic nanostructures of gold, copper, silver, carbon nanotubes, and graphene NPs (Boca et al., 2011; Yavuz et al., 2009). Several other morphologies of the IONPs have further been deployed for the photothermal cure of diseases. In a similar study Espinosa et al., utilizing nanocubes of IONPs, demonstrated and explained that during the exposure to the magnetic field and infrared laser irradiation nearby, much more heat (up to 3–5 fold) was produced than using by magnetic field alone. Further, it was investigated that the dual-mode stimulation produced sufficient heat by incorporating a lesser iron concentration (0.25 M), and laser irradiation of (0.3 W/cm^2), resulting in killing the cells completely, and thereby suppressing tumor (Espinosa et al., 2016).

Previous studies illustrate that heating might lead to irreversible damage to cellular proteins, hence they regulate apoptosis in tissues and cells (Mazario et al., 2017). A similar study conducted by Shen et al. (2015) developed a magnetic nanocluster for PTT with the help of near-infrared light irradiation. Laser irradiation also demonstrated total tumor regression in approximately three weeks in comparison to the control process. Researchers suggested that increment in PTT was because of the large crystalline and preferred lattice plane orientations of procured HCIONPs with reference to the normal Fe_3O_4 NPs. Figure 8.4 shows the morphology, crystalline and magnetic characterization of the synthesized Fe_3O_4 NPs.

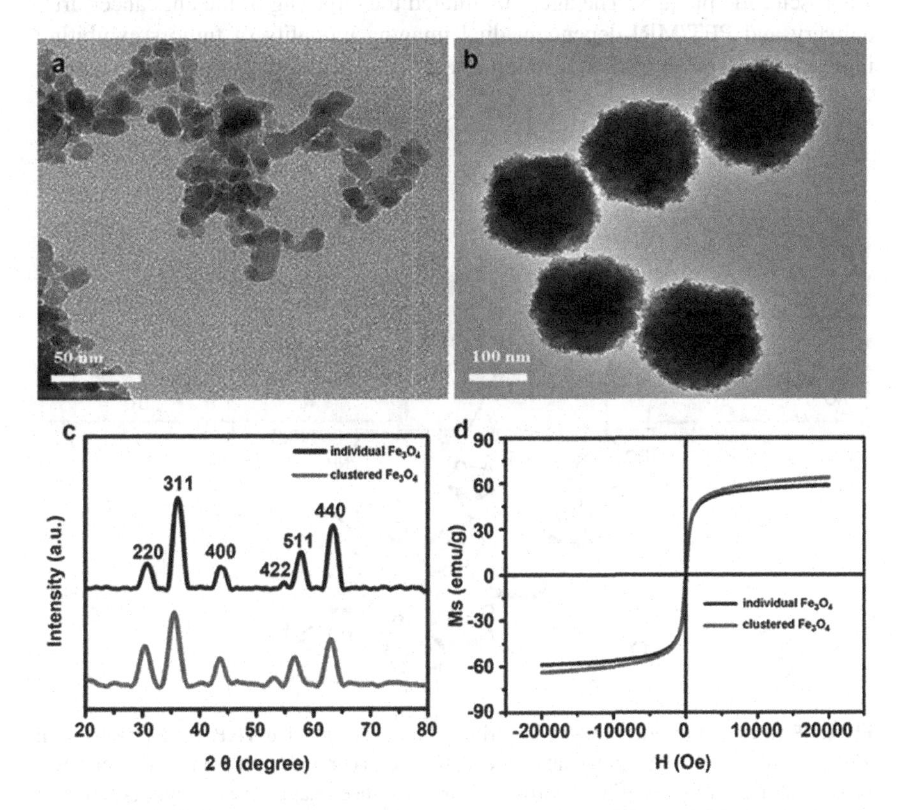

FIGURE 8.4 TEM images of (a) individual magnetic Fe_3O_4 NPs and (b) clustered magnetic Fe_3O_4 NPs. (c) The XRD pattern of individual and clustered magnetic Fe_3O_4 NPs. (d) Magnetization loops of individual and clustered magnetic Fe_3O_4 NPs. [Reproduced with permission from Shen et al. (2015) © Elsevier].

8.5 ROLE IN POSITRON EMISSION TOMOGRAPHIC IMAGING

PET serves as a nuclear imaging strategy capable of scanning the complete body and performing checks on the functioning of the tissues and the organs, thereby enabling quantification along with the activity of localization. Moreover, it is not used for anatomical or morphological imaging. Hence, MRI, CT, or US in combination as a hybrid system like PET/MRI or PET/CT demonstrate provision for higher magnification and cells anatomy along with the tissues (Evertsson et al., 2017). Further, a novel NPs system via bifunctional chelator dithiocarbamate–bisphosphonate conjugation to ^{64}Cu and dextran coated IONPs was developed for PET and MR imaging. Later, the labeling available for clinically approved IONPs along the ^{64}Cu-based dual functional chelator was done, and further the bimodality imaging was shown in vivo in lymph nodes. Nahrendorf et al. (2008) later interfused a PET tracer ^{64}Cu to magneto fluorescent NPs coated with dextran to extract a tri-modality reporter (^{64}Cu-TNP) for MRI, PET and fluorescence imaging. The multimodal NPs were found capable enough to sense the macrophages in atherosclerotic plaques. The agent facilitated the targeting of the anticancer drug delivery and PET/MRI-dependent dual imaging modality of tumors exhibiting integrin $\alpha_v\beta_3$. The simplified technique was found to be advantageous in forming

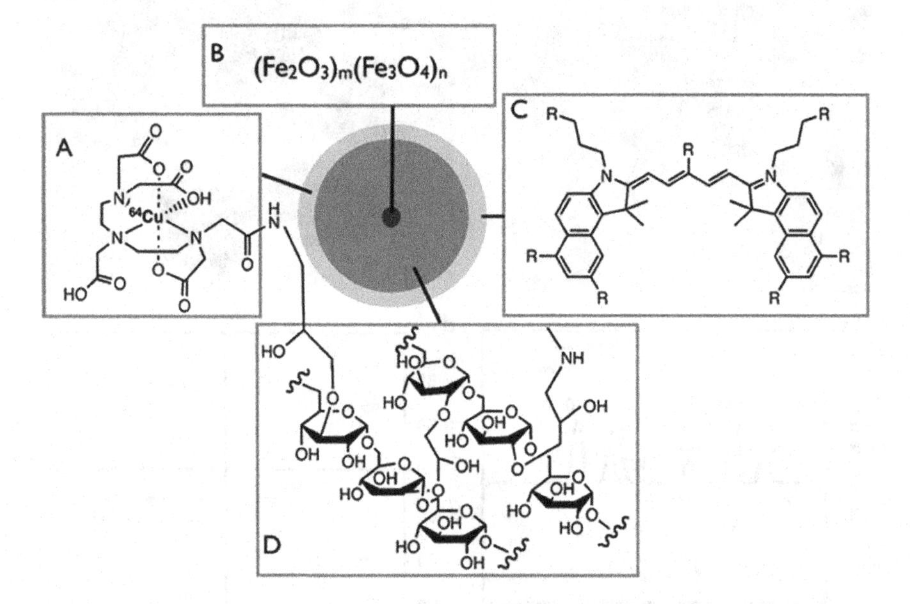

FIGURE 8.5 Schematic view of the trimodality reporter ^{64}Cu-TNP. (a) Derivatization with the chelator DTPA allows attachment of radiotracer ^{64}Cu. (b) Iron oxide core provides contrast in MRI. (c) Fluorochrome for fluorescence imaging, including fluorescence microscopy, flow cytometry, and fluorescence-mediated tomography. (d) Crosslinked aminated polysaccharide coating provides biocompatibility, determines blood half-life, and provides linker for attachment of tracers and potentially affinity ligands. [Reproduced with permission from Nahrendorf et al. (2008) © American Heart Association].

intrinsic radio-labeled NPs with chelators required. Additionally, these NPs were found to be useful for simultaneous PET and MRI imaging (Figure 8.5).

8.6 ROLE OF Fe_3O_4 NANOPARTICLES IN TREATMENT OF HUMAN BONE MARROW CELLS AND MOLECULAR CELL BINDING

8.6.1 BONE MARROW CELLS

The different cell populations of iron oxide nanoparticles are extracted from bone marrow cells which show various effects on G-CSF or MSCs or BMCs. The cure involving IONPs accelerates functional and morphological progress which may be more effective in curing the bone marrow cells, where the bone marrow cell cure and effect of the NPs lead to focus on the target area. In a similar study by Tumovcova et al. (2009), it was demonstrated that human mesenchymal stromal cells can be effective tools for future clinical application, however, their implication demands fast cell expansion in culture media applicable for clinical use. This encouraged them to study the effect of different culture media on development of colony, population doubling time, cycle of the cell and surface marker expression. The selection of the used serum affects the hMSC expansion and cell characteristics; α-MEM supplemented with hABS appears as an effective contender for clinical use.

The action of Resovist labeled bone marrow stem cells was done by Guo et al. (2013) in vivo, along with the cerebral transplantation in a model of Parkinson's disease applied on rats with the help of the MRI, the repercussions and fate of transplanted cells in different disease models along with the probability of higher time of observation. The IONPs have more impact in expanding new approaches of gene therapy based on the cells. Radu et al. (2015) studied the consequences of various anthropogenic and natural sources where they cited that the crystalline size of α-Fe_2O_3 NPs has better impact, and is capable of damaging the lung cells. The composition of hematite (α-Fe2O3) NPs at the time of peroxidation of lipid was 12.5 μg/mL. An antioxidative method in MRC-5 lung fibroblast cell's vulnerability was performed for one to three days. It was found that subjection to α-Fe_2O_3 NPs raised the lipid peroxidation by variable percentages, like 189%, 110%, and 81% after one day, two days, and three days, respectively, and they stated that MRC-5 antioxidant defense system is not capable of counteracting the oxidative stress caused by subjection to hematite NPs and may possibly demonstrate some impairment effects in human lung cells. Further, in a different study Horak et al. (2007) prepared surface-modified IONPs by precipitation of Fe(III) and Fe(II) salts along with ammonium hydroxide, and made comparison analysis of the obtained results with other methods. For the first method, a solution of D-mannose was used, whereas the other process included oxidation of sodium hypochlorite, and later mixing with the solution of D-mannose. Furthermore, the bone marrow stromal cells (rBMSCs) of the rat were marked with uncoated and IONPs modified using D-mannose and later with Endorem. Later, comparison was made between the efficiency of labeling and the viability of human and rat mesenchymal stem cells marked with Endorem, poly(L-lysine) (PLL)-altered Endorem, uncoated γ-Fe_2O_3, D-mannose and poly(N,N-dimethylacrylamide) (PDMAAm)-coated iron oxide nanoparticles. Studies suggest that magnetic iron oxide coated nanoparticles labeled cells were more effective than the Endorem. They showed rise in relaxation rates followed by visible contrast in in vitro magnetic resonance imaging of cells labeled along with the nanoparticles deposited at the surface, unlike the unlabeled cells.

8.6.2 Cell Labeling

Such study attracts researchers due to the physical characteristics and magnetic behaviors of iron oxide similar to the semiconductors along with the superparamagnetic IONPs. In their study, Mohammed et al. (2017) demonstrated that maghemite NPs exhibit the most potential applications in the fields of the environment and medicine. They also added that magnetic NPs are yet to achieve their optimal safety as compared to the hematite and efficiency considering the challenges for use in combination in vivo. The satisfactory results were acquired by controlling shortcomings with the study in advancement of magnetic aimed mediators for preclinical trials. Darroudi et al. (2014) later demonstrated a dose-dependent toxicity along with the nontoxic impact of composition less than 62.5 μg/mL for in vitro cytotoxicity on neuro2A cells. It was later discovered that higher composition of intracellular IONPs act more efficiently to cell physiology in a composition dependent form. Higher concentrations of iron oxide cores effect the actin formation, development, and cytoskeleton of focal adhesion complexes and may also alter the protein expression levels. In another study it was validated that the implication of intracytoplasmic iron oxide in labeled cells was confirmed due to the adaptation of simple Prussian blue staining. The investigation involves the decline in uptake beyond the detection threshold along with the Prussian blue stain. The HeLa cells were cultured in presence of ferumoxide-PLL and MION-46L-PLL with variable iron compositions, and found that the time, along with the cells, influenced huge amounts of intracytoplasmic Prussian blue-positive particles in the cells marked with ferumoxides contrary to the one labeled with MION-46L-PLL; further, the cells were cocultured for a period of six hours, followed by washing, and later were assessed by cell labeling using Prussian blue stain. The amalgamation of FDA-permitted dextran-coated iron oxide MR contrast agents along with the tannic acids (TA) were tested, resulting obtaining ferumoxide-TA and MION- 46L-TA complexes permitting a unique and nonspecific process of magnetically labeling the stem cells. A similar study came as a major breakthrough for understanding the in vivo trafficking of transplanted cells using MRI of superparamagnetic iron oxide-labeled cells. The method of cellular diagnosis using MRI is basically believed to be less sensitive in comparison to techniques like positron emission tomography, single photon emission-computed tomography or optical fluorescence microscopy. The impacts of IONPs on cell homeostasis needs a multidisciplinary perspective, considering several parameters as

IONPs may cause a range of toxicological belongings, and being applied as a model system involving several other toxicological investigations of IONPs. A similar study demonstrated the application of amino-superparamagnetic IONPs as an MRI contrast carrier and dual functional labeling probe for the neural stem cells.

Henceforth, it can be inferred that the dedicated transfer of iron oxide NPs along with the DOX plays an important role in increasing the cell labeling performance and cytotoxic activity. It has been suggested that in future nanodrug delivery systems, iron oxide NPs can be enhanced for several applications and are useful in cell labeling treatment.

8.6.3 Targeted Drug Delivery

Traditional methods of cancer treatment include surgery, chemotherapy, radiotherapy, and combinational therapy to alleviate the tumors. However, damage to cells surrounding a tumor and radio-resistance of cells limits the effectiveness of these conventional therapies. The multifunctional nature of IONPs has been recently exploited for the effective drug delivery for treatment of cancer and other diseases. Islamian et al. (2017) used superparamagnetic mesoporous hydroxyapatite conjugated doxorubicin and deoxy-D-glucose nanocomposites to boost breast cancer chemo and radiotherapy. They worked on SKBR3 and T47D breast cancer cell culture models and reported that the cell viability was significantly decreased with combined nanocomposite effect compared to the radiotherapy alone. Mechanistically, they found that targeting of breast cancer cells achieved with deoxy-d-glucose moiety conjugated on IONPs surface and doxorubicin acted as therapeutic agent, thus the combined effect led toimproved breast cancer radiotherapy by increased localization of NPs. When investigated further, it was found that the extent of cytotoxicity in tumor cells was much more, with minimum side effects and damage to normal healthy cells.

In another strategy, Ye et al. (2017) suggested that Fe_3O_4 NPs can increase the efficacy of cryoablation; a process that uses extreme cold conditions to treat cancerous cells. Their data indicated that Fe_3O_4 NPs altered intracellular ice formation ability during freezing, recrystallization, and thawing, which leads to the enhanced killing of MCF-7 cells. Therefore, the idea of enhanced ablation using IONPs can be successfully applied to effectively treat tumors in near future (Figure 8.6).

A few of the selective biological treatments using iron oxide NPs are elaborated in this section. As per the studies conducted by Faust et al. (2014), the impact of the different sized α-Fe_2O_3 NPs using human placentas in vitro model was shown, where they assumed that the NPs diameter had a severe impact on the workability of the epithelium. They found that the epithelium is affected in several ways with the exposure of different diameters. The study also suggested that the α-Fe_2O_3 NPs of diameters 50 nm and 78 nm disturbed the epithelium and apoptosis's junctional coalition. Contrarily, α-Fe_2O_3 NPs having a diameter as small as 15 nm had minimal impact on the epithelium. The study illustrates that more

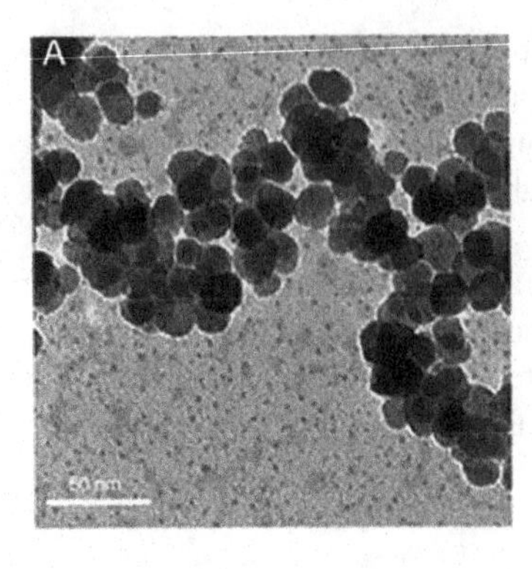

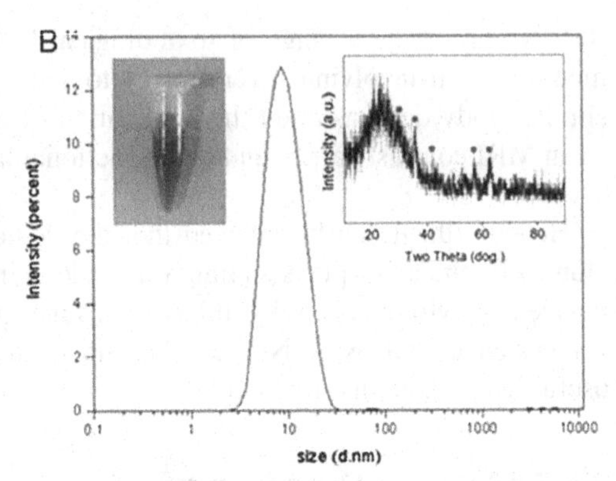

FIGURE 8.6 (a) TEM and (b) DLS images of the Fe_3O_4 nanoparticles. The inserted images (left) and (right) represent Fe_3O_4 nanoparticle aqueous solution and XRD image of Fe_3O_4 nanoparticle, respectively. The scale bar is 50 nm [Reproduced from Ye et al. 2017].

changes are experienced with NPs that have large diameters as compared to those with small diameters. From the reported study it can be inferred that the fetal/maternal link provided shuttling nutrients to the placenta, which provided hindrance to the mother's immune parts, and allowing exchange of waste and gas, coupling disruption or affecting these cells during pregnancy, or even causing fetal malnutrition. Correia Carreira et al. (2015) illustrated the transport and uptake of IONPs occurring in vitro in the BeWo b30 placental barrier model, and they discovered that the size of the NPs altering the transport and toxicity was not particularly impactful in this study for 25 nm and 50 nm. From the study it can be inferred that recommendations could be made for the parameters inducing toxicity and the placental barrier model for more appropriate extrapolation to certain unique biological applications. Several modifiers including glycol chitosan, amino agarose, leaf extracts, arginine, DMSA, dextran, saccharide, L-arginine, DMAA, D-mannose, covalent binding, polymer, silane, ethylene glycol, citric acid, gelatin, PVP, c(RGDyK), polysaccharide, PVA, PEI, PEG, PVC, peptide, and tannins have also been used successfully for several biological applications (Bakhtiary et al., 2016; Rajendran et al., 2011, Marcu et al., 2013; Calero et al., 2015). Proper investigations to understand the connectivity between the size of the nanoparticles and the structures of iron oxide or varying morphologies play a crucial role in disease detection in the early stages and post-treatments.

REFERENCES

Bakhtiary, Z., Saei, A.A., Hajipour, M.J., Raoufi, M., Vermesh, O., Mahmoudi, M. 2016. Targeted superparamagnetic iron oxide nanoparticles for early detection of cancer: Possibilities and challenges. *Nanomedicine* 12 (2), 287–307.

Boca, S.C., Potara, M., Gabudean, A.-M., Juhem, A., Baldeck, P.L., Astilean, S. 2011. Chitosan-coated triangular silver nanoparticles as a novel class of biocompatible, highly effective photothermal transducers for in vitro cancer cell therapy. *Cancer Lett.*, 311 (2), 131–140.

Calero, M., Chiappi, M., Lazaro-Carrillo, A., Rodríguez, M.J., Chichón, F.J., Crosbie-Staunton, K., Prina-Mello, A., Volkov, Y., Villanueva, A., Carrascosa, J.L. 2015. Characterization of interaction of magnetic nanoparticles with breast cancer cells. *J. Nanobiotechnol.* 13 (16), 1–15.

Correia Carreira, S., Walker, L., Paul, K., Saunders, M. 2015. The toxicity, transport and uptake of nanoparticles in the in vitro BeWo b30 placental cell barrier model used within nanoTEST. *Nanotoxicology* 9, 66–78.

Darroudi, M., Hakimi, M., Goodarzi, E., Oskuee, R.K. 2014. Superparamagnetic iron oxide nanoparticles (SPIONs): Green preparation, characterization and their cytotoxicity effects. *Ceram. Int.* 40 (9), 14641–14645.

Espinosa, A., Di Corato, R., Kolosnjaj, T.J., Flaud, P., Pellegrino, T., Wilhelm, C. 2016. Duality of iron oxide nanoparticles in cancer therapy: Amplification of heating efficiency by magnetic hyperthermia and photothermal bimodal treatment. *ACS Nano* 10 (2), 2436–2446.

Evertsson, M., Kjellman, P., Cinthio, M. Andersson, R., Tran, T.A., Zandt, R., Grafstrom, G., Toftevall, H., Fredriksson, S., Ingvar, C., Strand, S.-E., Jansson, T. 2017. Combined magnetomotive ultrasound, PET/CT, and MR imaging of 68Ga-labelled superparamagnetic iron oxide nanoparticles in rat sentinel lymph nodes in vivo. *Sci. Rep.* 7, 4824.

Faust, J.J., Zhang, W., Chen, Y., Capco, D.G. 2014. Alpha-Fe_2O_3 elicits diameter-dependent effects during exposure to an in vitro model of the human placenta. *Cell Bio. Toxic.* 30 (1), 31–53.

Guo, L., Ding, W., Zheng, L.-M. 2013. Synthesis and evaluation of c(RGDyK)-coupled superparamagnetic iron oxide nanoparticles for specific delivery of large amount of doxorubicin to tumor cell. *J. Nanopart. Res.* 15, 1720.

Hope, M.D., Hope, T.A., Zhu, C., Faraji, F., Haraldsson, H., Ordovas, K.G., Saloner, D. 2015. Vascular imaging with ferumoxytol as a contrast agent. *AJR Am. J. Roentgenol.* 205 (3), 366–373.

Horak, D., Babic, M., Jendelová, P., Herynek, V., Trchová, M., Pientka, Z., Pollert, E., Hájek, M., Syková, E. 2007. D-mannose-modified iron oxide nanoparticles for stem cell labeling. *Bioconjug. Chem.* 18 (3), 635–644.

Huang, D.M., Hsiao, J.K., Chen, Y.C., Chien, L.Y., Yao, M., Chen, Y.K., Ko, B.S., Hsu, S.C., Tai, L.A., Cheng, H.Y., Wang, S.W., Yang, C.S., Chen, Y.C. 2009. The promotion of human mesenchymal stem cell proliferation by superparamagnetic iron oxide nanoparticles, *Biomaterials* 30 (22), 3645–3651.

Huber, D.L. 2005. Synthesis, properties, and applications of iron nanoparticles. *Small* 1 (5), 482–501.

Islamian, J.P., Hatamian, M., Aval, N.A., Rashidi, M.R., Mesbahi, A., Mohammadzadeh, M., Jafarabadi, M.A. 2017. Targeted superparamagnetic nanoparticles coated with 2-deoxy-d-glucose and doxorubicin more sensitize breast cancer cells to ionizing radiation. *Breast* 33, 97–103.

Ju, Y., Zhang, H., Yu, J., Tong, S., Tian, N., Wang, Z., Wang, X., Su, X., Chu, X., Lin, J., Ding, Y., Li, J., Sheng, F., Hou, Y. 2017. Monodisperse Au–Fe_2C Janus nanoparticles: An attractive multifunctional material for triple-modal imaging-guided tumor photothermal therapy. *ACS Nano* 11 (9), 9239–9248.

Karimzadeh, I., Aghazadeh, M., Doroudi, T., Ganjali, M.R., Kolivand, P.H. 2017. Superparamagnetic iron oxide (Fe_3O_4) nanoparticles coated with PEG/PEI for biomedical applications: A facile and scalable preparation route based on the cathodic electrochemical deposition method. *Adv. Phys. Chem.* 2017, 9437487.

Kolosnjaj, T.J., Wilhelm, C. 2017. Magnetic nanoparticles in cancer therapy: How can thermal approaches help? *Nanomedicine* 12, 573–575.

Lu, Y.-J., Lin, P.-Y., Huang, P.-H., Kuo, C.-Y., Shalumon, K., Chen, M.-Y., Chen, J.-P., 2018. Magnetic graphene oxide for dual targeted delivery of doxorubicin and photothermal therapy. *Nanomaterials* 8, 193.

Marcu, A., Pop, S., Dumitrache, F., Mocanu, M., Niculite, C.M., Gherghiceanu, M., Lungu, C., Fleaca, C.P., Ianchis, R., Barbut, A., Grigoriu, C., Morjan, I. 2013. Magnetic iron oxide nanoparticles as drug delivery system in breast cancer. *Appl. Surf. Sci.* 281, 60–65.

Mazario, E., Forget, A., Belkahla, H., Lomas, J.S., Decorse, P., Biraud, A.C. Verbeke, P., Wilhelm, C., Ammar, S., El Hage Chahine, J.-M., Hemadi, M. 2017. Functionalization of iron oxide nanoparticles with HSA protein for thermal therapy. *IEEE Trans. Magn.* 53 (11), 1–5.

Mohammed, L., Gomaa, H.G., Ragab, D., Zhu, J. 2017. Magnetic nanoparticles for environmental and biomedical applications: A review. *Particuology* 30 (1), 1–14.

Nahrendorf, M., Zhang, H., Hembrador, S., Panizzi, P., Sosnovik, D.E., Aikawa, E., Libby, P., Swirski, F.K., Weissleder, R. 2008. Nanoparticle PET-CT imaging of macrophages in inflammatory atherosclerosis. *Circulation* 117 (3), 379–387.

Nehra, P., Chauhan, R.P., Garg, N., Verma, K. 2018. Antibacterial and antifungal activity of chitosan coated iron oxide nanoparticles. *Br. J. Biomed. Sci.* 75 (1), 13–18.

Patra, J.K., Ali, M.S., Oh, I.-G., Baek, K.-H. 2017. Proteasome inhibitory, antioxidant, and synergistic antibacterial and anticandidal activity of green biosynthesized magnetic Fe_3O_4 nanoparticles using the aqueous extract of corn (*Zea mays* L.) ear leaves. *Artif. Cells Nanomed. Biotechnol.* 45 (2), 349–356.

Radu, M., Dinu, D., Sima, C., Burlacu, R., Hermenean, A., Ardelean, A., Dinischiotu, A. 2015. Magnetite nanoparticles induced adaptive mechanisms counteract cell death in human pulmonary fibroblasts. *Toxicol In Vitro.* 29 (7), 1492–1502.

Rajendran, K., Karunagaran, V., Mahanty, B., Sen, S. 2011. Biosynthesis of hematite nanoparticles and its cytotoxic effect on HepG2 cancer cells. *Int. J. Biol. Macromol.* 74, 376–381.

Shen, S., Wang, S., Zheng, R., Zhu, X., Jiang, X., Fu, D., Yang, W. 2015. Magnetic nanoparticle clusters for photothermal therapy with near-infrared irradiation. *Biomaterials* 39, 67–74.

Singh, S. 2017. Glucose decorated gold nanoclusters: A membrane potential independent fluorescence probe for rapid identification of cancer cells expressing Glut receptors. *Colloids Surf. B Biointerfaces* 155, 25–34.

Sohn, C.H., Park, S.P., Choi, S.H., Park, S.H., Kim, S., Xu, L., Kim, S.H., Hur, J.A., Choi, J., Choi, T.H. 2015. MRI molecular imaging using GLUT1 antibody-Fe_3O_4 nanoparticles in the hemangioma animal model for differentiating infantile hemangioma from vascular malformation. *Nanomedicine* 11 (1), 127–135.

Turnovcova, K., Ruzickova, K., Vanecek, V., Sykova, E., Jendelova, P. 2009. Properties and growth of human bone marrow mesenchymal stromal cells cultivated in different media. *Cytotherapy* 11 (7), 874–885.

Wu, W., Wu, Z., Yu, T., Jiang, C., Kim, W.-S. 2015. Recent progress on magnetic iron oxide nanoparticles: Synthesis, surface functional strategies and biomedical applications. *Sci. Technol. Adv. Mater.* 16 (2), 23501.

Xie, J., Chen, K., Huang, J., Lee, S., Wang, J., Gao, J., Li, X., Chen, X. 2010. PET/NIRF/MRI triple functional iron oxide nanoparticles. *Biomaterials* 31 (11), 3016–3022.

Yavuz, M.S., Cheng, Y., Chen, J., Cobley, C.M., Zhang, Q., Rycenga, M., Xie, J., Kim, C., Song, K.H., Schwartz, A.G., Wang, L.V., Xia, Y. 2009. Gold nanocages covered by smart polymers for controlled release with near-infrared light. *Nat. Mat.* 8 (12), 935–939.

Ye, P., Kong, Y., Chen, X., Li, W., Liu, D., Xie, Y., Zhou, Y., Zou, H., Chang, Z., Dai, H., Kong, X., Liu, P. 2017. Fe_3O_4 nanoparticles and cryoablation enhance ice crystal formation to improve the efficiency of killing breast cancer cells. *Oncotarget* 8 (7), 11389–11399.

Yew, Y.P., Shameli, K., Miyake, M., Khairudin, N.B.A., Mohamad, S.E., Naiki, T., Lee, K.X. 2020. Green biosynthesis of superparamagnetic magnetite Fe_3O_4 nanoparticles and biomedical applications in targeted anticancer drug delivery system: A review. *Arab. J. Chem.* 13 (1), 2287–2308.

9 Iron-based Catalysis toward Biomass Processing

9.1 INTRODUCTION

The development of science and technologies for the replacement of fossil resources with renewable biomass has become a high priority interest due to climate change concerns and dwindling fossil resources. Structurally the biomass-based molecules are very different from the carbon forms in fossil resources, and the conversion of these structurally diverse, highly oxygenated systems to renewable equivalents is a formidable challenge. The transition metal-based energy efficient catalysis processes with high turnover numbers, reasonable reaction rates, and efficient recyclability are the most sought-after methods in tackling these challenges. Many catalysis chemistry methods developed in synthetic organic chemistry can be adopted for biobased systems, and iron-based catalysis has attracted recent attention in this field for many reasons, including the following:

- Iron is the second most abundant element in earth's crust with ~5% abundance; the metal, its oxides and many of its salts are readily available and inexpensive (Plietker, 2011).
- Iron chemistry is well established; many compounds can be prepared using well-known, simple techniques. Several forms of naturally occurring oxides or ores can be used as catalysts without any modifications.
- The magnetic properties of iron can be used to recover the catalyst for recycling.
- Relatively nontoxic, currently iron is considered by the regulatory authorities as a "metal with minimum safety concern"; 1300 ppm residual iron is deemed acceptable in drug substances (Egorova and Ananikov, 2016).
- Iron is centrally located in the d-block of elements with formal oxidation states ranging from –II to + VI and hence useful in reductive and oxidative manifolds alike.
- Lewis acidity of iron varies from fairly modest to very high allowing a wide range of applications in Lewis acid catalyzed reactions.
- Traditional iron-based coal liquefaction catalysts like limonite and hematite iron ore may have promising features for the catalytic conversions of bio-derived carbon feedstocks in reforming and cracking processes as well (He et al., 2016).

DOI: 10.1201/9781003243632-9

9.2 IRON CATALYZED THERMAL LIQUEFACTION OF BIOMASS

The thermal liquefaction or direct liquefaction of biomass is the thermochemical conversion of biomass into a liquid product using water, an organic medium, or solvent mixtures. In comparison with pyrolysis using dry biomass, direct liquefaction usually runs at a relatively lower temperature, takes less energy, and leads to low tar yields, thus attracting more attention in recent years (Gollakota et al., 2018). The use of iron-based catalysts in coal liquefaction is known over a century. The original 1913 direct coal liquefaction method known as Bergius process was developed using iron sulfide as the catalyst. In this process finely divided coal and catalyst were mixed with heavy oil and hydrogenated at 400°C–500°C and at a hydrogen pressure of 50–70 MPa to produce hydrocarbons. The application of iron-based catalysts in biomass liquefaction can be seen as an adoption of this classic technology for renewable biomass resources.

9.2.1 IRON CATALYZED LIQUEFACTION IN WATER

Iron catalysis has been studied since the 1980s in the liquefaction of woody biomass (Bestue-Labazuy et al., 1985). In a more recent work, Sun et al. studied the liquefaction of paulownia wood in hot-compressed water to produce heavy oil and water-soluble oil at 180°C–360°C, using iron catalysts, and discovered that iron promoted the liquefaction producing oil in 36.34% yield at 340°C (Sun et al., 2011). The same group reported a comparison of catalytic activities of iron and Na_2CO_3 as well, where iron contributed to the maximum production of heavy oil, while Na_2CO_3 led to the minimization of the solid residue (Sun et al., 2010). Then, in similar experiments, Xu and Lad (2008) studied the heavy oil production from jack pine wood in sub-/near-critical water using a series of catalysts: $Ca(OH)_2$, $Ba(OH)_2$, and $FeSO_4$. However, in this instance the $FeSO_4$ addition resulted in lower heavy oil yields in comparison to the other two catalysts.

In another example of iron catalyzed liquefaction of woody biomass where oak wood was liquefied in water the highest biocrude oil yield reported was 40% (de Caprariis et al., 2019). Miyata and coworkers studied the liquefactions of softer biomass forms such as oil palm empty fruit bunches in water at 300°C. In this detailed study they quantified various compounds in the bio-oils and explored the mechanisms of iron catalysis (Miyata et al., 2017, 2018). During this work, they found that iron may assist in the conversion of unstable intermediates into stable oil compounds, while reducing the char production. Miyata et al. have proposed plausible reaction pathways for the Fe-assisted liquefaction of biomass. According to their explanation, at first cellulose and hemicellulose are hydrolyzed to C_5 and C_6 sugars, and then these sugars are cleaved into unstable C_2–C_4 aldehydes through FeO-assisted retro-aldol condensation. Later on, most aldehydes are converted to stable alcohols through the hydrogenation, catalyzed by Fe(0), and the rest of the aldehydes are condensed into insoluble polymers. In addition they have proposed three possible reaction pathways for Fe(0) catalyzed C=O to CH–OH reduction as well (Miyata et al., 2018). In another recent case, wheat stalk was used as the biomass and the catalysts were: Na_2CO_3, Fe and Na_2CO_3-Fe mixture. In these experiments the Na_2CO_3-Fe mixture demonstrated the best performance in biomass conversion into bio-oil at a relatively low temperature (Chen et al., 2019).

9.2.2 Iron Catalyzed Liquefaction in Organic Solvents

Direct coal liquefactions are generally carried out in organic solvents such as heavy oil, and the original process used FeS as the catalyst (Ali and Zhao, 2020). Following this century-old process a few researchers have tested the use of iron-based compounds such as $Fe(CO)_5$–S and FeS for liquefactions in organic solvents. In an early example in the 1990s, liquefaction of spirulina, a high-protein microalgae, was conducted in tetralin (TL), 1-methylnaphthalene (1-MN) or toluene (TOL) using Fe $(CO)_5$–S as catalyst at 300°C–425°C in H_2 or CO environments producing bio-oil in moderate yields (Matsui et al., 1997). Later, liquefaction of the same microalgae was investigated using $FeSO_4$ or FeS as catalyst in sub- and supercritical ethanol, with a catalyst dosage 5–7 wt% producing over 50% yields of bio-oils (Huang et al., 2011). In another example, liquefaction of lipid-extracted microalgae was carried out in water or iso-propanol using waste steel furnace residue as the catalyst. In this instance the researchers reported significant reductions in oxygen and nitrogen contents in the bio-oil product due to the addition of the iron catalyst (Wagner et al., 2018). The hydro-liquefaction of jack pine wood has been carried out in sub- and supercritical ethanol with and without $FeSO_4$/FeS, and it was found that 5 wt % $FeSO_4$ is more efficient than 5 wt % FeS as a catalyst (Xu and Etcheverry, 2008). In liquefaction in ethanol rice husk was also treated in sub- and supercritical ethanol to produce bio-oil with catalysts $FeSO_4$, FeS, Na_2CO_3, and NaOH. However, the researchers found only a slight increase bio-oil yields with the addition of $FeSO_4$ or FeS in comparison to the uncatalyzed experiment (Huang et al., 2013). In addition, Kim et al. (2017) reported a catalytic coprocessing of biomass with conventional heavy oil by using Fe_3O_4 in supercritical m-xylene and n-dodecane to produce light oil in high yields.

9.2.3 Iron Catalyzed Liquefaction in Water-organic Solvent Mixtures

Iron catalyzed liquefaction in water-organic solvent mixtures is another interesting approach as the aqueous phase allows better solubility of the salts. Mottweiler et al. (2015) have studied the catalytic activity of $FeCl_3$-derived iron complexes in oxidative cleavage of lignin model compounds in DMSO – water mixtures at 82°C and 100°C. The lignin model erythro-dilignol could be efficiently hydrolyzed using Fe-DABCO catalyst in 97% yield. In a similar approach, Li et al. focused on the investigation of reductive cleavage of the C–O bond in lignin by using lignin model compounds with nitrogen-doped carbon-supported iron catalysts; phenol yields of 95% and toluene yield of 90% were achieved from the hydrogenolysis of α-O-4 lignin model compounds (Li et al., 2018).

Furthermore, Nakasaka et al. (2017) reported a two-step lignin degradation to produce phenols from dealkaline lignin using solid acid and iron oxide catalysts in water-1-butanol mixtures. In the first step, lignin was depolymerized to n-hexane, ethyl acetate soluble fractions, and another heavy fraction. In the second step, fractionated liquid could be separated into phenols, polymer additives, coke, and residue. Hematite iron ore is an inexpensive iron catalyst and the co-liquefaction of lignin and lignite coal in water-ethanol mixtures were

TABLE 9.1
Selected Examples of Research Progress in Iron Catalyzed Liquefactions of Biomass in Water, Organic Solvents and Water-Organic Solvent Mixtures

Biomass	Catalyst	Solvent	Liquefaction Condition	Major Findings
Birchwood	$FeSO_4$	Water	260–320°C, Stirred autoclave	~40% yield with $FeSO_4$
Oak wood	Fe, Fe_3O_4, Fe_2O_3	Water	260–320°C, Stirred autoclave	~40% yield with Fe, Fe_3O_4 was magnetically recovered
Oil palm empty fruit bunches	Iron	Water	N_2, 300°C, Hastelloy C, high-pressure reactor	Bio-oil 79% yield at a H_2O: biomass ratio of 5:1
Wheat stalk	Na_2CO_3, Fe, Na_2CO_3 +Fe	Water	N_2, 210–270°C, 2–8 MPa	Heavy bio-oil 24.25% yield with Na_2CO_3+Fe
Anchusa-azurea	$Na_2B_4O_7$, $FeCl_3$	Methanol, iso-propanol	260, 280 and 300°C, stainless steel autoclave	$FeCl_3$ is more active at 260°C in iso-propanol and at 280°C in methanol
Micro-crystalline cellulose	Fe_3O_4	Tetralin, decalin, m-xylene	H_2, 0.1–3.5 MPa, 300–450°C, stainless steel reactor with stirring	Cellulose conversion 98.3%. Yield in tetralin 96.7%, yield in decalin 93.1%
Hardwood organosolv lignin	Fenton catalyst (Fe^{3+} and H_2O_2)	Ethanol	N_2, 250°C, 7 MPa, in a flask with stirring	Bio-oil 66% yield
Sweet sorghum bagasse	Fenton catalyst (Fe^{3+} and H_2O_2)	Ethanol	N_2, 250°C, 6.5 MPa, Parr stirring reactor	Phenolic oil 75.8% yield, Fenton catalyst benefits phenolic oil production and enzymatic hydrolysis

Source: [Reproduced with permission from Du et al. (2020) © Elsevier].

conducted to produce heavy oil in 40% yield (Paysepar et al., 2018). The hydrogen-reduced hematite resulted in a slightly better 42% yield of heavy oil under similar conditions. Later, the same group investigated the effects of solvents (water, ethanol, and mixed ethanol-water and the addition of formic acid), lignite/lignin ratio, and iron ore catalyst loading and reached the highest heavy oil yield of 48.7% in ethanol with 10 (v) % formic acid and H2-reduced hematite ore catalyst (Shui et al., 2019). In addition, higher bio-oil yields are reported in the liquefaction of corn stalk in water-ethanol systems with H2-reduced hematite ore in comparison to hematite as the catalyst. In another example of using alcohol-water solvent, Wu et al. prepared a bifunctional Ni–HPMo/Fe_3O_4@Al-MCM-41 catalyst and used it in the liquefaction of switchgrass in water-iso-butanol mixtures (Wu et al., 2019). These experiments could achieve biomass conversion of 84.7%, and a liquid yield of 55.0%. Further examples of research progress in iron catalyzed liquefactions of biomass in water, organic solvents, and water-organic solvent mixtures are shown in Table 9.1.

9.3 PYROLYSIS OF BIOMASS THROUGH IRON CATALYST

Pyrolysis is a thermal decomposition of biomass occurring in the absence of oxygen. The main products of biomass pyrolysis are bio-oil, bio-char and gases including methane, hydrogen, carbon monoxide and carbon dioxide. Depending on the pyrolysis temperature, environment, and the mass transfer rate, pyrolysis can be used to produce bio-oil or gaseous product mixtures (Kan et al., 2016). In fast pyrolysis, biomass is rapidly heated to 450°C–600°C in the absence of air. Under these conditions, organic vapors, pyrolysis gases, and charcoal are produced. The vapors are condensed to bio-oil and typically 60–75 wt % of the feedstock can be converted into bio-oil by this technique. The pyrolysis in a gasifier at 800°C–1000°C can be used to produce a gas mixture composed of mainly carbon monoxide and hydrogen known as syngas. The pyrolysis method for the production of bio-oils as well as syngas are some of the most intensively studied fields in the biomass based renewable energy arena in the past 30 years. There are many advantages of this process such as the variety of biomass forms that can be used as feedstocks. Then, there are disadvantages such as high energy cost, high water content, instability and complexity of bio-oils, as well as difficulties in upgrading to high energy fuels. The addition of metals as catalysts is one of the most sought-after approaches to improving the quality of the oil and lowering the pyrolysis temperature.

9.3.1 Iron Catalyzed Conventional Pyrolysis of Biomass to Produce Bio-oil and Syngas

The common application of iron-assisted conventional pyrolysis yields bio-oil and syngas; this topic is discussed under this section of the review. The use of iron oxides or iron ores as pyrolysis catalysts is a very attractive proposition since these forms are inexpensive, and other forms of iron may also get oxidized by oxygen in biomass at high pyrolysis temperatures (Lu et al., 2020). In fact, many researchers have tested the use of iron ores and iron oxides as catalysts in pyrolysis. For example, Khelfa et al. (2009) reported a steam gasification of *Miscanthus giganteus* at 850°C by using hematite (Fe_2O_3) as the catalyst. The best product yield of 94.8 wt% gases and 5.2 wt% liquids were achieved in a 20 minute pyrolysis.

Limonite is another iron ore popular as a pyrolysis catalyst; it consists of a mixture of hydrated iron (III) oxide – hydroxide in varying compositions. The generic formula is often written as $FeO(OH)_n$. H_2O, and limonite is one of the three principal iron ores, the others being hematite and magnetite. In an application of limonite ore, a pyrolysis of lignins has produced high-quality bio-oils containing 55% alkyl phenolics and 27% aromatics, with the use of limonite as catalyst (Hita et al., 2018). Similarly, Virginie et al. (2012) have compared the catalytic activity of iron containing olivine to the original olivine in the pyrolysis of pine wood and found that the tar yield could be significantly reduced by using iron-olivine. Furthermore, they have noted that iron-olivine not only served as a catalyst for tar reduction, but also acted as oxygen carrier during the biomass gasification. In addition, Wei et al. (2017) compared the gasification of lignin pellets with, and without, the addition of iron ore and found that the gaseous yield with the catalyst was much higher than the conventional biomass gasification in the temperature range of 900°C–1060°C.

9.3.2 Iron Catalyzed Microwave Pyrolysis of Biomass

Microwave-assisted pyrolysis is a relatively new technique among various pyrolysis technologies. Compared to conventional heating, microwave heating has the advantage of leading to more uniform temperature distribution within the sample when an appropriate microwave absorber is blended with the sample. The leading work on catalytic microwave- assisted pyrolysis of biomass was carried out by Wan et al. (2009) using various catalysts including metal oxides, salts, and acids on the biomass forms corn stover and aspen. In a follow-up work, Zhang et al. reported the char-supported metal catalyzed microwave pyrolysis of rice husk in the production of high-quality syngas (Zhang et al., 2015). Their investigations indicated that metal ions impregnated into the char improved the microwave absorption capacity of biomass, and that both Ni and Fe catalysts could enhance the conversion of tar into syngas. They further focused on the investigation of the effects of iron (III) catalysts on the pyrolysis of moso bamboo after impregnating onto activated carbon (Dong et al., 2018). Interestingly, the catalyst improved the energy efficiency of bamboo pyrolysis, promoted the yields of H_2 and CO, and suppressed the formation of CH_4 and CO_2. In another example, fast, catalyst-assisted microwave gasification of corn stover was conducted with Al_2O_3 supported Fe, Co, Ni catalysts, and the catalysts efficiently increased the gas production and improved the gas quality (Xie et al., 2014).

Similarly, Bartoli et al. (2016) have reported the pyrolysis of α-cellulose in a multimode microwave oven using different microwave absorbers such as iron, SiC, SiO_2, Al_2O_3, and carbon powder. In this instance, the highest bio-oil yield of 37.6% was achieved when iron was used as the catalyst. Furthermore, the iron could be recovered from the bio-char with a magnet for reuse. In order to study the production of bio-oil from municipal solid waste, Suriapparao and Vinu (2015) have carried out a catalytic microwave pyrolysis of mixtures of cellulose, paraffin oil, kitchen waste and garden waste, which served as a model for municipal solid waste.

9.3.3 Iron Catalyzed Pyrolysis of Biomass to Produce Graphite

Iron based catalysts have been used in the direct pyrolytic transformation of raw biomass into nanostructured graphitic carbon. In this context, Thompson et al. (2015) reported the use of Fe_3C nanoparticles fabricated from the thermal decomposition of an iron precursor and the carbothermal reduction. These iron-carbon nanoparticles were used in catalytic graphitization of softwood sawdust, where they claimed that the catalyst helped to generate intertwined graphitic tubules at relatively low temperatures of around 800°C. Furthermore, the meso-porosity of the produced graphite was found to be dependent on the iron content.

In another recent example, two iron-based Lewis acid catalysts $FeCl_2$ and $FeCl_3$ were used in the hydrothermal carbonization of α-cellulose (Abd Hamid et al., 2015). In comparison to the graphite produced without a catalyst, both iron chloride catalysts enhanced the surface area of the produced carbon nanomaterials with the complete carbonization of α-cellulose at 200°C. The graphite formed with the $FeCl_2$ catalyst possessed a greater C=O functionality than the product using $FeCl_3$.

9.4 BIOBASED HYDROGEN PRODUCTION UTILIZING IRON CATALYST

Hydrogen has a high energy content of about 130 MJ/kg and is a clean energy carrier with the potential to replace fossil fuels. Hydrogen is already in use as an experimental fuel in buses and public transport systems in a few major cities around the world. Hydrogen fuel cell powered electric cars are also available in selected markets as a test program (Gurz et al., 2017). However, it uses commercial methods for its production that are not environmentally friendly; they require a major energy input and entail high costs. On the other hand, biohydrogen production offers an environmentally friendly alternative. It makes good use of organic wastes and requires less energy, and in particular the microorganism-based methods require relatively low energy input compared to thermochemical and electrolysis processes. The use of iron-based catalysts is known in a number of areas of biobased hydrogen production, such as the use of iron-based catalysts in biomass pyrolysis to produce hydrogen-rich syngas and the use of iron-based enzymes in biochemical hydrogen production methods.

9.4.1 Thermo-chemical Methods of Hydrogen Production Using Iron Catalysts

The latest advances in hydrogen production by thermocatalytic conversion of biomass have been recently reviewed by Li et al. (2019). This brief but up-to-date review summarizes the use of various transition metals and noble metals in enhancing the hydrogen yield in syngas. The use of iron has caught the attention of many researchers due to its abundance. The metal catalysts were generally supported on a catalyst support such as carbon, silica, alumina, or zeolite (Li et al., 2019). The Ni-based catalysts have superior properties for the production of hydrogen. However, when iron is added as the cocatalyst, the synergy effect can further improve the catalytic activity. In Ni–Fe catalysts, Ni and Fe are known to have strong metal-support interaction, such that both the particle size and the carbon deposition rate can be decreased. The strong synergistic effect of Ni–Fe bimetallic system found in some hydrogen production experiments is probably due to the presence of a number of reducible metal oxide species in these catalysts (Yao et al., 2017). The use of Ni and Fe combination deposited on γ-Al_2O_3 nano-catalysts support also showed similar results in bagasse gasification processes where significant improvements in total gas and H_2 production yields are reported (Jafarian et al., 2017). The inexpensive Fe_2O_3 iron oxide was widely used in many trials, as the catalyst in hydrogen production through biomass gasification. The addition of another metal oxide or mixed metal oxides is known to improve the performance of iron oxides. In one example, Duman et al. (2014) tested a 1:1 mixture of CeO_2 and Fe_2O_3, demonstrating excellent hydrogen production yields and low tar formation during the steam gasification of safflower seed cake biomass. The iron naturally found in the biomass can also act as an effective catalyst in the biomass pyrolysis. In fact, Skoulou and Zabaniotou investigated this effect in biomass pyrolysis (Skoulou and Zabaniotou, 2012). Their study showed that olive kernel pyrolysis char is a highly effective catalyst in hydrogen production in comparison to cellulosic biomass char, due to its porous structure, increased surface area, and high iron content. Iron-based catalysts are known for their ability in tar depletion as well, further, these catalysts in their metallic states are generally known to exhibit a better tar-cracking capacity than their corresponding oxides.

9.4.2 Biochemical Methods of Hydrogen Production Using Iron Containing Enzyme Catalysts

Biological hydrogen production is based on activity of specific class of enzymes called hydrogenases. These are the key enzymes involved in the metabolism of molecular H_2 and are generally grouped into three subgroups based on the metal cofactor present at the active site, namely [Fe–Fe], [Ni–Fe], and [Fe] hydrogenases. Typical Fe containing active sites of hydrogenase enzymes are shown in Figure 9.1. On the whole, hydrogenases catalyze the reversible reduction of protons to molecular hydrogen as shown in the equation (Stephenson and Stickland, 1931).

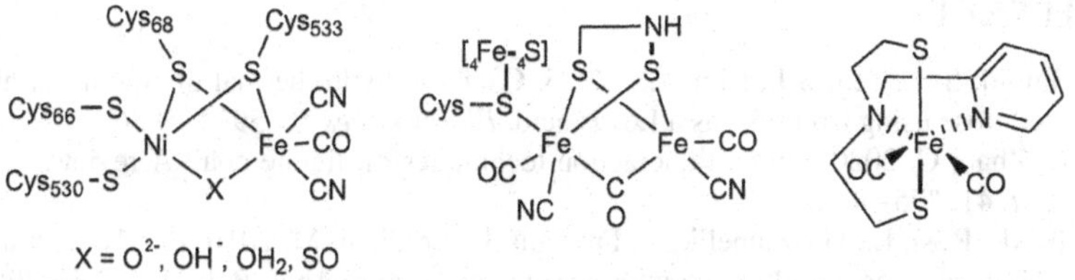

FIGURE 9.1 Typical Fe containing active sites of [Ni–Fe], [Fe–Fe], and [Fe] hydrogenase enzymes [Reproduced with permission from Kim and Kim (2011) © Elsevier].

The biochemistry of hydrogenase enzymes as well as related microbial systems is of interest to a number of disciplines. These hydrogenase enzymes are sensitive to molecular oxygen since O_2 can also bind to the Fe in the active site destroying the activity of the enzyme (Bingham et al., 2012). Oxygen deactivation is one of the major limitations for applying this technology for large-scale hydrogen production. In general, [Fe–Fe] hydrogenases prefer proton reduction to molecular hydrogen, therefore most studies on biohydrogen production have used this class of hydrogenases (Das et al., 2006). The current interest in biohydrogen production using biochemical processes and various hydrogenase systems is evident from the number of review articles published in this area in the last couple of years (Lin et al., 2018). Enzymatic biohydrogen production can be achieved by two different processes: light-dependent, or light-independent. The light independent process is known as dark fermentation as well, and this process does not require energy from light. Of the two, the dark fermentation process is more attractive due to its simplicity of operation, relatively high hydrogen conversion, flexibility in cultivation, and the possibility of the use of organic waste as the carbon source (Singh and Wahid, 2015). Even though dark fermentation is generally more productive, there are many efforts to improve hydrogen production through photo-fermentation as well. One common approach is the improvement of biohydrogen yield through enhancement in light conversion efficiency as it will facilitate photo-fermentation.

One of the straightforward ways to advance photo-fermentation is by adding a suitable chemical enhancer. A number of studies have shown that addition of certain chemicals, such as ethylenediaminetetraacetic acid (EDTA), iron, or molybdenum salts and vitamins can increase the biohydrogen production rates and yields by significant amounts (Budiman and Wu, 2018). The use of iron-based compounds such as iron oxides or iron oxide nanoparticles to enhance the biochemical process is an interesting development in biohydrogen production. Malik et al. (2014) studied the kinetics of nanocatalyzed dark fermentative biohydrogen production from molasses-based distillery wastewater with the addition of iron oxide nanoparticle (10–200 mgL^{-1}) to the wastewater to enhance the biohydrogen production. At the optimum conditions, the maximum rate of hydrogen production and specific hydrogen yield with the iron oxide nanoparticles were found to be 80.7 mL/h and 44.28 mL H_2/g COD (chemical oxygen demand).

REFERENCES

Abd Hamid, S.B., Teh, S.J., Lim, Y.S. 2015. Catalytic hydrothermal upgrading of alpha-cellulose using iron salts as a Lewis acid. *Bioresources* 10, 5974–5986.

Ali, A., Zhao, C. 2020. Direct liquefaction techniques on lignite coal: A review. *Chin. J Cat.* 41, 375–389.

Bartoli, M., Rosi, L., Giovannelli, A., Frediani, P., Frediani, M. 2016. Pyrolysis of alpha-cellulose using a multimode microwave oven. *J. Anal Appl. Pyro.* 120, 284–296.

Bestue-Labazuy, C., Soyer, N., Bruneau, C., Brault, A. 1985. Wood liquefaction with hydrogen or helium in the presence of iron additives. *Can. J. Chem. Eng.* 63, 634–638.

Bingham, A.S., Smith, P.R., Swartz, J.R. 2012. Evolution of an [FeFe] hydrogenase with decreased oxygen sensitivity. *Int. J. Hyd. Energy* 37, 2965–2976.

Budiman, P.M., Wu, T.Y. 2018. Role of chemicals addition in affecting biohydrogen production through photofermentation. *Energy Conver. Manag.* 165, 509–527.

Chen, Y.X., Cao, X.D., Zhu, S., Tian, F.S., Xu, Y.Y., Zhu, C.S., et al. 2019. Synergistic hydrothermal liquefaction of wheat stalk with homogeneous and heterogeneous catalyst at low temperature. *Bio. Technol.* 278, 92–98.

Das, D., Dutta, T., Nath, K., Kotay, S.M., Das, A.K., Veziroglu, T.N. 2006. Role of Fe-hydrogenase in biological hydrogen production. *Current Sci. Bangalore* 90, 1627.

de Caprariis, B., Bavasso, I., Bracciale, M.P., Damizia, M., De Filippis, P., Scarsella, M. 2019. Enhanced biocrude yield and quality by reductive hydrothermal liquefaction of oak wood biomass: effect of iron addition. *J. Anal. Appl. Pyro.* 139, 123–130.

Dong, Q., Niu, M.M., Bi, D.M., Liu, W.Y., Gu, X.X., Lu, C. 2018. Microwave-assisted catalytic pyrolysis of moso bamboo for high syngas production. *Bioresour. Technol.* 256, 145–151.

Du, H., Deng, F., Kommalapati, R.R., Amarasekara, A.S. 2020. Iron-based catalysts in biomass processing. *Renew. Sustain. Energy Rev.* 134, 110292.

Duman, G., Watanabe, T., Uddin, M.A., Yanik, J. 2014. Steam gasification of safflower seed cake and catalytic tar decomposition over ceria modified iron oxide catalysts. *Fuel Process Technol.* 126, 276–283.

Egorova, K.S., Ananikov, V.P. 2016. Which metals are green for catalysis? Comparison of the toxicities of Ni, Cu, Fe, Pd, Pt, Rh, and Au salts. *Angew Chem. Int. Ed.* 55, 12150–12162.

Gollakota, A.R.K., Kishore, N., Gu, S. 2018. A review on hydrothermal liquefaction of biomass. *Renew. Sustain. Energy Rev.* 81, 1378–1392.

Gurz, M., Baltacioglu, E., Hames, Y., Kaya, K. 2017. The meeting of hydrogen and automotive: A review. *Int. J. Hyd. Energy* 42, 23334–23346.

He, L., Li, S., Lin, W. 2016. Catalytic cracking of pyrolytic vapors of low-rank coal over limonite ore. *Energy Fuel* 30, 6984–6990.

Hita, I., Heeres, H.J., Deuss, P.J. 2018. Insight into structure-reactivity relationships for the iron-catalyzed hydrotreatment of technical lignins. *Bioresour. Technol.* 267, 93–101.

Huang, H.J., Yuan, X.Z., Zeng, G.M., Liu, Y., Li, H., Yin, J., et al. 2013. Thermochemical liquefaction of rice husk for bio-oil production with sub- and supercritical ethanol as solvent. *J. Anal Appl. Pyro.* 102, 60–67.

Huang, H.J., Yuan, X.Z., Zeng, G.M., Wang, J.Y., Li, H., Zhou, C.F., et al. 2011. Thermochemical liquefaction characteristics of microalgae in sub- and supercritical ethanol. *Fuel Process Technol.* 92, 147–153.

Jafarian, S., Tavasoli, A., Karimi, A., Norouzi, O. 2017. Steam reforming of bagasse to hydrogen and synthesis gas using ruthenium promoted NiFe/γAl_2O_3 nano-catalysts. *Int. J. Hyd. Energy* 42, 5505–5512.

Kan, T., Strezov, V., Evans, T.J. 2016. Lignocellulosic biomass pyrolysis: a review of product properties and effects of pyrolysis parameters. *Renew. Sustain. Energy Rev.* 57, 1126–1140.

Khelfa, A., Sharypov, V., Finqueneisel, G., Weber, J.V. 2009. Catalytic pyrolysis and gasification of Miscanthus Giganteus: Haematite (Fe_2O_3) a versatile catalyst. *J. Anal. Appl. Pyrol.* 84, 84–88.

Kim, D.W., Koriakin, A., Jeong, S.Y., Lee, C.H. 2017. Co-processing of heavy oil with wood biomass using supercritical m-xylene and n-dodecane solvents. *Kor. J. Chem. Eng.* 34, 1961–1969.

Kim, D.-H., Kim, M.-S. 2011. Hydrogenases for biological hydrogen production. *Bioresour. Technol.* 102, 8423–8431.

Li, J., Sun, H., Liu, J.X., Zhang, J.J., Li, Z.X., Fu, Y. 2018. Selective reductive cleavage of C-O bond in lignin model compounds over nitrogen-doped carbon-supported iron catalysts. *Mol. Cat.* 452, 36–45.

Li, S., Zheng, H., Zheng, Y., Tian, J., Jing, T., Chang, J.-S., et al. 2019. Recent advances in hydrogen production by thermo-catalytic conversion of biomass. *Int. J. Hyd. Energy* 44, 14266–14278.

Lin, C.-Y., Nguyen, T.M.-L., Chu, C.-Y., Leu, H.-J., Lay, C.-H. 2018. Fermentative biohydrogen production and its byproducts: A mini review of current technology developments. *Renew. Sustain. Energy Rev.* 82, 4215–4220.

Lu, Q., Li, W., Zhang, X., Liu, Z., Cao, Q., Xie, X., et al. 2020. Experimental study on catalytic pyrolysis of biomass over a Ni/Ca-promoted Fe catalyst. *Fuel* 263, 116690.

Malik, S.N., Pugalenthi, V., Vaidya, A.N., Ghosh, P.C., Mudliar, S.N. 2014. Kinetics of nano-catalysed dark fermentative hydrogen production from distillery wastewater. *Energy Procedia.* 54, 417–430.

Matsui, T.-O., Nishihara, A., Ueda, C., Ohtsuki, M., Ikenaga, N.-O., Suzuki, T. 1997. Liquefaction of micro-algae with iron catalyst. *Fuel* 76, 1043–1048.

Miyata, Y., Sagata, K., Hirose, M., Yamazaki, Y., Nishimura, A., Okuda, N., et al. 2017. Fe-assisted hydrothermal liquefaction of lignocellulosic biomass for producing high-grade bio-oil. *ACS Sustain. Chem. Eng.* 5, 3562–3569.

Miyata, Y., Sagata, K., Yamazaki, Y., Teramura, H., Hirano, Y., Ogino, C., et al. 2018. Mechanism of the Fe-assisted hydrothermal liquefaction of lignocellulosic biomass. *Ind. Eng. Chem. Res.* 57, 14870–14877.

Mottweiler, J., Rinesch, T., Besson, C., Buendia, J., Bolm, C. 2015. Iron-catalysed oxidative cleavage of lignin and beta-O-4 lignin model compounds with peroxides in DMSO. *Green Chem.* 17, 5001–5008.

Nakasaka, Y., Yoshikawa, T., Kawamata, Y., Tago, T., Sato, S., Takanohashi, T., et al. 2017. Fractionation of degraded lignin by using a water/1-butanol mixture with a solid-acid catalyst: A potential source of phenolic compounds. *Chem. Cat. Chem.* 9, 2875–2880.

Paysepar, H., Ren, S.B., Kang, S.G., Shui, H.F., Xu, C.B. 2018. Catalytic co-liquefaction of lignin and lignite coal for aromatic liquid fuels and chemicals in mixed solvent of ethanol-water in the presence of a hematite ore. *J. Anal. Appl. Pyro.* 134, 301–308.

Plietker, B. 2011. *Iron catalysis: Fundamentals and applications.* Berlin Heidelberg: Springer Science & Business Media.

Shui, H.F., Zou, D.H., Wu, H.H., He, F., Wang, X.L., Pan, C.X., et al. 2019. Co-liquefaction of Xilinguole lignite and lignin in ethanol/water solvents under a cheap iron ore catalyst. *Fuel* 251, 629–635.

Singh, L., Wahid, Z.A. 2015. Methods for enhancing biohydrogen production from biological process: A review. *J. Ind. Eng. Chem.* 21, 70–80.

Skoulou, V., Zabaniotou, A. 2012. Fe catalysis for lignocellulosic biomass conversion to fuels and materials via thermochemical processes. *Cat. Today* 196, 56–66.

Stephenson, M., Stickland, L.H. 1931. Hydrogenase: A bacterial enzyme activating molecular hydrogen: The properties of the enzyme. *Biochem. J.* 25, 205.

Sun, P.Q., Heng, M.X., Sun, S.H., Chen, J.W. 2010. Direct liquefaction of paulownia in hot compressed water: Influence of catalysts. *Energy* 35, 5421–5429.

Sun, P.Q., Heng, M.X., Sun, S.H., Chen, J.W. 2011. Analysis of liquid and solid products from liquefaction of paulownia in hot-compressed water. *Energy Conv. Manag.* 52, 924–933.

Suriapparao, D.V., Vinu, R. 2015. Bio-oil production via catalytic microwave pyrolysis of model municipal solid waste component mixtures. *RSC Adv.* 5, 57619–57631.

Thompson, E., Danks, A.E., Bourgeois, L., Schnepp, Z. 2015. Iron-catalyzed graphitization of biomass. *Green Chem.* 17, 551–556.

Virginie, M., Adanez, J., Courson, C., de Diego, L.F., Garcia-Labiano, F., Niznansky, D., et al. 2012. Effect of Fe-olivine on the tar content during biomass gasification in a dual fluidized bed. *Appl. Cat. B Environ.* 121, 214–222.

Wagner, J.L., Perin, J., Coelho, R.S., Ting, V.P., Chuck, C.J., Franco, T.T. 2018. Hydrothermal conversion of lipid-extracted microalgae hydrolysate in the presence of isopropanol and steel furnace residues. *Waste Biomass Val.* 9, 1867–1879.

Wan, Y.Q., Chen, P., Zhang, B., Yang, C.Y., Liu, Y.H., Lin, X.Y., et al. 2009. Microwave-assisted pyrolysis of biomass: Catalysts to improve product selectivity. *J. Anal. Appl. Pyro.* 86, 161–167.

Wei, R.F., Feng, S.H., Long, H.M., Li, J.X., Yuan, Z.S., Cang, D.Q., et al. 2017. Coupled biomass (lignin) gasification and iron ore reduction: A novel approach for biomass conversion and application. *Energy* 140, 406–414.

Wu, H.T., Zheng, J.L., Wang, G.Q. 2019. Catalytic liquefaction of switchgrass in isobutanol/water system for bio-oil development over bifunctional Ni-HPMo/Fe_3O_4@ Al-MCM-41 catalysts. *Renew. Energy* 141, 96–106.

Xie, Q.L., Borges, F.C., Cheng, Y.L., Wan, Y.Q., Li, Y., Lin, X.Y., et al. 2014. Fast microwave-assisted catalytic gasification of biomass for syngas production and tar removal. *Bioresour Technol.* 156, 291–296.

Xu, C.B., Etcheverry, T. 2008. Hydro-liquefaction of woody biomass in sub- and super-critical ethanol with iron-based catalysts. *Fuel* 87, 335–345.

Xu, C.B., Lad, N. 2008. Production of heavy oils with high caloric values by direct liquefaction of woody biomass in sub/near-critical water. *Energy Fuel* 22, 635–642.

Yao, D., Wu, C., Yang, H., Zhang, Y., Nahil, M.A., Chen, Y., et al. 2017. Co-production of hydrogen and carbon nanotubes from catalytic pyrolysis of waste plastics on Ni-Fe bimetallic catalyst. *Energy Conv. Manag.* 148, 692–700.

Zhang, S.P., Dong, Q., Zhang, L., Xiong, Y.Q. 2015. High quality syngas production from microwave pyrolysis of rice husk with char-supported metallic catalysts. *Bioresour Technol.* 191, 17–23.

10 Environmental Hazard and Toxicity Study of Iron-based Nanomaterials

10.1 TOXIC EFFECT ON MAMMALIAN NERVE CELLS

Phenrat et al. (2009) reported the effects of age and surface modification on the toxicity behavior of nanoscale zero-valent iron particles (nZVI). The study utilized tests on mammalian cells by fresh nZVI, aged nZVI (nearly a year old), magnetite, and polyaspartate surface-modified nZVI. Various properties of nanoparticles such as "redox" activity, agglomeration, and sedimentation rate were found to impose morphological changes on neuron cells and rodent microglia of mammalian cells However, capping of surface-modified nZVI by polyaspartate, imposed less toxicity behavior due to limited exposure of the iron particles to the cells. A remarkable impact was observed for fresh nZVI, whereas insignificant morphological changes were induced by aged nZVI in mitochondrial cells, and in neuron cells the ATP levels were reduced.

In the presence of a polymeric coating, toxicity effects are very limited or even absent (Li et al., 2010). All forms of nZVI aggregated in soil and water in the presence of a high concentration of calcium ions, and thus, addition of calcium salts may help in reducing the toxicity of groundwater due to nZVI (Keller et al., 2012).

To overcome such toxicological issues various research studies were carried out to advance and modify the production of green nanomaterials, which are less toxic and thus more eco-friendly for engineered nanoparticles (Shakibaie et al., 2013; Usha Rani and Rajasekharreddy, 2011).

Nadagouda et al. (2010) investigated the toxic effects of biosynthesized nZVI on human keratinocyte cells. Figures 10.1 and 10.2 show the transmission electron microscopy (TEM) images of the synthesized Fe NPs. Green tea extract and borohydride were utilized for synthesizing nZVI. The biocompatibility was assessed for 24–48 h using an assay of methyl tetrazolium (MTS) and lactate dehydrogenase (LDH), which were exposed to cell lines. Due to higher average particle diameter LDH leakage increased inducing stress on the cellular membrane. Hence, green synthesized nZVI was utilized owing to much smaller particle size and non-toxic effects on human keratinocytes compared to chemically synthesized nZVI. Table 10.1 shows the various zero-valent nanoscale iron particles prepared from tea extract in the study.

DOI: 10.1201/9781003243632-10

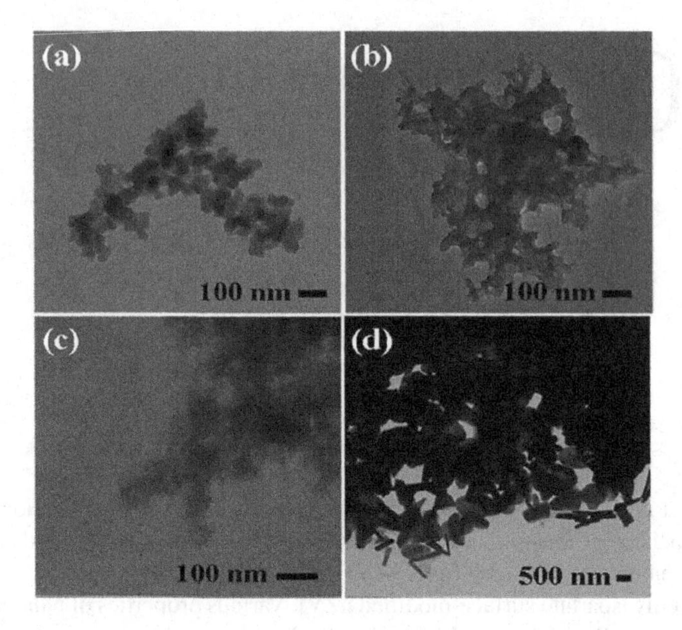

FIGURE 10.1 TEM images of (a) T1, (b) T2, (c) T3, and (d) T4 samples. [Reproduced from Nadagouda et al. (2010) © Royal Society of Chemistry].

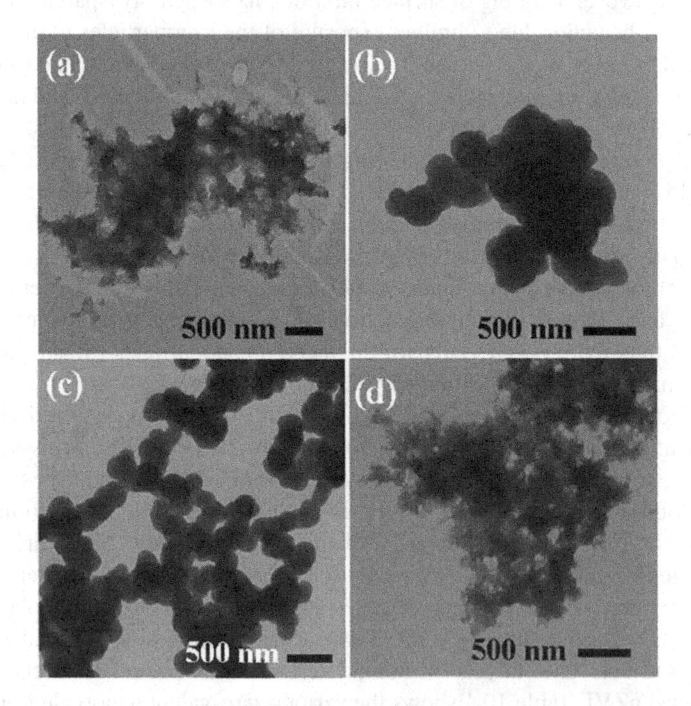

FIGURE 10.2 TEM images of (a) T5, (b) T6, (c) T7, and (d) T8 samples. [Reproduced from Nadagouda et al. (2010) © Royal Society of Chemistry].

TABLE 10.1
Preparation of Nanoscale Zero-valent Iron (nZVI) Particles Using Tea

Sample Code	Description
T1	10 mL tea extract + 1 mL 0.1 N $Fe(NO_3)_3$ solution
T2	5 mL tea extract + 5 mL 0.1 N $Fe(NO_3)_3$ solution
T3	1 mL tea extract + 5 mL 0.1 N $Fe(NO_3)_3$ solution
T4	1 mL tea extract + 10 mL 0.1 N $Fe(NO_3)_3$ solution
T5	5 mL tea extract + 4 mL 0.1 N $Fe(NO_3)_3$ solution
T6	5 mL epicatechin (0.01 N) extract + 1 mL 0.1 N $Fe(NO_3)_3$
T7	4 mL epicatechin (0.01 N) extract + 4 mL 0.1 N $Fe(NO_3)_3$
T8	5 mL tea extract + 2 mL 0.1 N $Fe(NO_3)_3$ solution

10.2 TOXIC EFFECT ON AQUATIC LIFE

Zebrafish embryos were employed for testing the toxicity effect of prepared nanoparticles in varying dosage. The studies confirmed developmental impairment and DNA damage of the zebrafish embryo. An increased mortality rate along with slower beat and hatching rate was noticed with increasing nanoparticle concentration in zebrafish eggs. The adverse effect of high concentration maghemite nanoparticles resulted in larvae malformation and suggested harm to aquaculture.

The toxic effect of iron in three different forms was studied by Chen et al. (2013). The study employed solutions of nZVI coated with carboxymethyl cellulose (CMC-nZVI), magnetite (Fe_3O_4), and ferrous ion solution Fe(II). Figure 10.3 shows TEM images of (a) carboxymethyl cellulose-stabilized nanoscale zero-valent iron (CMC-nZVI), and (b) nFe_3O_4. Such solutions were introduced to test toxicity effects on the early life stages of medaka fish. Comparable to magnetite

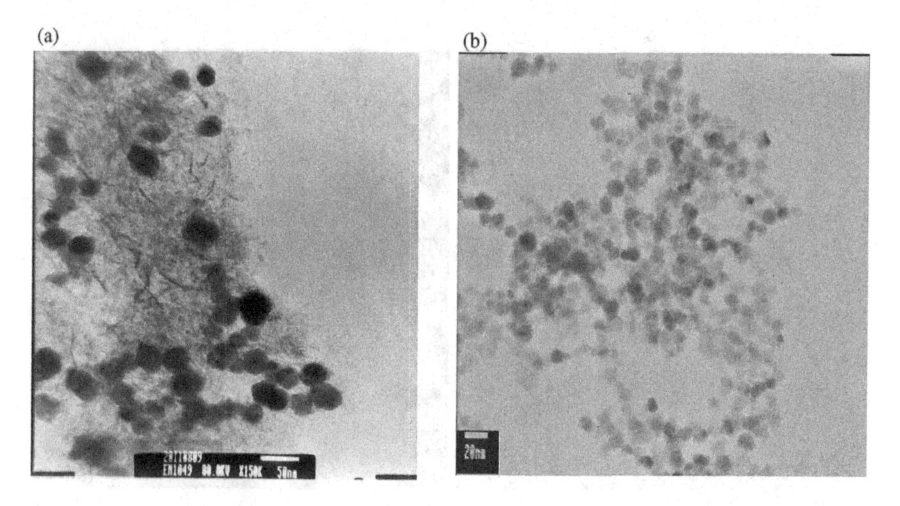

FIGURE 10.3 TEM images of (a) carboxymethyl cellulose-stabilized nanoscale zero-valent iron (CMC-nZVI), and (b) nFe_3O_4. [Reproduced with permission from Chen et al. (2013) © Elsevier].

and Fe(II) solution, CMC-nZVI was found to have a more toxic effect on the embryo. Due to the presence of various oxidized forms of iron in the CMC-nZVI solution, effects such as hypoxia, developmental toxicity, and reactive oxygen species (ROS) oxidative stress occurred in the embryos. The results conveyed how physicochemical properties such as chemical reactivity, the oxidation state, and particle aggregation of nZVI alter in the aqueous medium, which further affects the toxicity characteristics of nZVI in medaka fish.

The effects of nZVI on medaka fish (*Oryzias latipes*) and their embryos were investigated recently. A dose and time-dependent decrease in superoxide dismutase (SOD) and malondialdehyde (MDA) activities was observed in the embryos. A significant decrease of SOD and glutathione (GSH) activity was observed in liver and brain samples taken from the adults, but as the exposure time increased, the adults appeared to recover from the exposure by adjusting the levels of antioxidant enzymes (Li et al., 2009). The toxicity of iron oxide nanoparticles was found not to reduce through stabilization or capping agents in many studies.

Baumann et al. (2014) studied toxicity behavior with four different iron NPs coated with ascorbate, citrate, dextran, and polyvinylpyrrolidone. The toxicity tests were performed on the neonates of *Daphnia magna* (water flea). Ascorbate and dextran-coated iron NPs showed the highest immobilization, whereas citrate-coated NPs showed lower immobilization. Incomplete ecdysis was observed for acerbate-, dextran-, and citrate-coated iron NPs when present in high dosage, whereas polyvinylpyrrolidone-coated iron NPs showed a negligible negative effect. For other functionalized iron NPs, PVP-coated NPs showed the highest colloidal stability. Moreover, it was found that stabilizing forces controlling hydrodynamic diameter did not have any effect on toxicity, but factors like colloidal stability and release of ions induced ROS in daphnids.

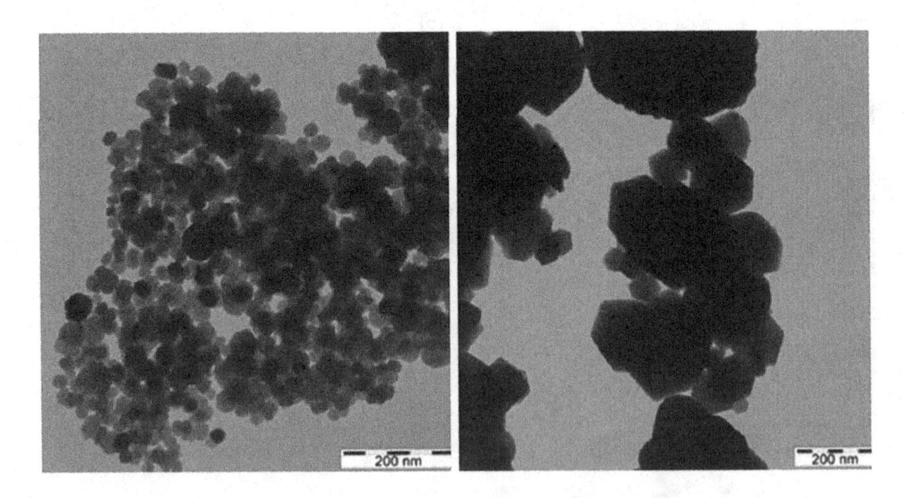

FIGURE 10.4 TEM image of nanosized (left) and bulk (right) Fe_3O_4 particles. [Reproduced with permission from Blinova et al. (2017) © Springer Nature].

The toxicity of nanosized and bulk iron oxide NPs was evaluated by Blinova et al. (2017) and tested on *Daphnia magna* (water flea) and *Lemna minor* (duckweed). Both the iron oxide NPs showed negligible biological effects with very low toxicity (EC_{50} < 100 ppm), but at magnetite concentrations of 10 and 100 ppm, a decrease in the number of neonates hatched for *Daphnia magna* was noticed. TEM images of the synthesized Fe NPs are shown in Figure 10.4.

10.3 TOXIC EFFECT ON MICROORGANISMS

Iron-based nanomaterials, despite their tremendous environmental applications, carry high risks for the environment. Improper management of wastewater and other refuse from industries could lead to specific environmental hazards in groundwater and soil. Synthesized iron nanomaterial for specific remediation purposes can get transported from one medium to another. Zero-valent iron nanoparticles (nZVI) are considered to be the most reactive of all iron nanoparticles. During the treatment of groundwater contaminants with zero-valent iron nanoparticles (nZVI), it was observed that they are transported through permeable soil and undergo transformation on interacting with contaminants as well with the exposed environment. The environment is directly or indirectly affected due to toxic impacts generated by iron NPs on microorganisms and soil fauna.

It was reported that the induced stress of nanosized iron exerts a toxic effect on soil microorganisms and causes an alteration in microbial biomass (Sacca et al., 2014). The nanosized iron particles interact strongly with enzymes of living bodies and metals in the environmental arena due to their unique morphological features, particle size, surface energy, self-assembly, and synthesis process. However, biosynthesized nanoparticles being deployed with harmful chemicals like hydrazine as capping agents makes them a mild toxic agent for the environment. Antisari et al. (2013) reported the change in microbial mass of the soil due to the toxic impacts of engineered iron nanoparticles. Moreover, Fajardo et al. (2013) studied the transcriptional and proteomic stress activity due to nano zero-valent iron on soil bacterium *Bacillus cereus*. Das et al. (2018), in a recent study utilized strain LS4- a *Desulfovibrio* bacterium for preparing maghemite nanoparticles (Fe_2O_3) from saltpan sediments, through in situ synthesis. An average particle diameter of 18 nm was obtained by the in vitro synthesis.

Auffan et al. (2008), through experimental studies, have drawn a direct relationship between the cytotoxicity and oxidation state of iron nanoparticles. Figure 10.5 shows the influence of the pH on the colloidal stability of nMaghemite, nMagnetite, and nZVI. The cytotoxic impacts of different iron oxides such as nZVI, magnetite, and maghemite were tested toward gram-negative bacteria *E. Coli* and were compared. From the tests it was found that nZVI had a higher toxicity effect as compared to other iron oxide NPs. Hence, the study suggested the oxidation state of iron nanoparticles played a vital role in toxicity. Such a phenomenon arises due to the generation of oxidative stress from ROS. ROS denotes highly unstable radicals such as superoxide radicals, and hydroxyl radicals which disrupt the functioning of cells on getting adsorbed over the cell membrane.

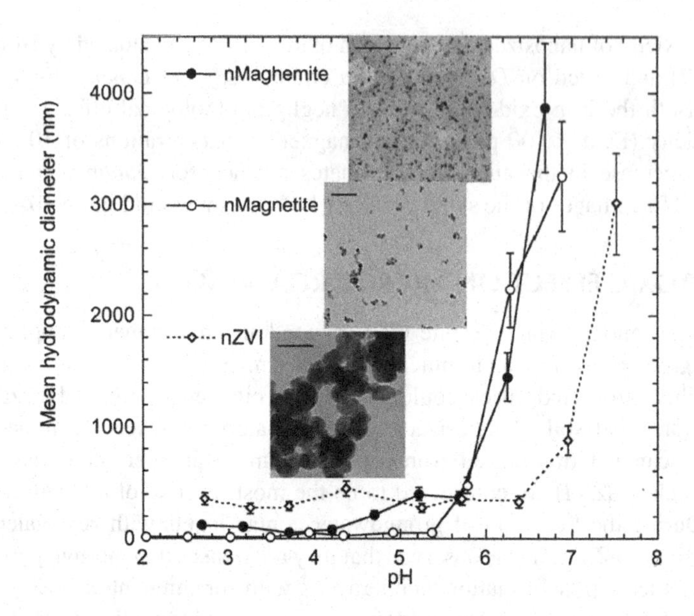

FIGURE 10.5 Influence of the pH on the colloidal stability of nMaghemite, nMagnetite, and nZVI (0.01 mol/L NaCl, 25°C, incubation time: 1 h). Inset pictures: transmission electron micrograph of nMaghemite, nMagnetite, and nZVI before contact with bacteria. Black dash: 90 nm scale [Reproduced with permission from Auffan et al. (2008) © American Chemical Society].

In another study Lee et al. (2008) reported strong bactericidal activity at anaerobic conditions exhibited by nZVI. The study revealed a linear correlation between the nZVI dose and the log inactivation of *E. coli*. Under aerobic conditions due to the oxidation phenomenon, the toxicity of nZVI significantly gets reduced due to the formation of an oxide layer. Li et al. (2010) confirmed the reduced bactericidal effects of nZVI due to complete oxidation in aerobic conditions. Further studies concluded that Fe(II) released from nZVI under anaerobic conditions was found to be more toxic. Hence, the studies revealed that physical disruption of the cell membrane was triggered by nZVI and led to physical damage whereas cell inactivation was enhanced by the biocidal effects of Fe(II).

Nanosized ZVI (nZVI) is found to be capable of removing viruses (e.g., MS-2) from water by inactivating them and/or irreversibly adsorbing the viruses to the iron (You et al., 2005). A study done by Lee et al. (2008) says that nZVI particles exhibited a bactericidal effect on *Escherichia coli* that was not observed with other types of iron-based compounds, such as iron oxide nanoparticles, microscale ZVI, and Fe^{3+} ions. Nanosized ZVI coated with humic acid showed minor toxicity to *E. coli* (Li et al., 2010).

Barhoumi and Dewez (2013) studied the toxicity effect of superparamagnetic iron oxide NPs toward green algae *Chlorella vulgaris*, since algae acts as an ecological indicator and provides the health of the aquatic ecosystem. The study

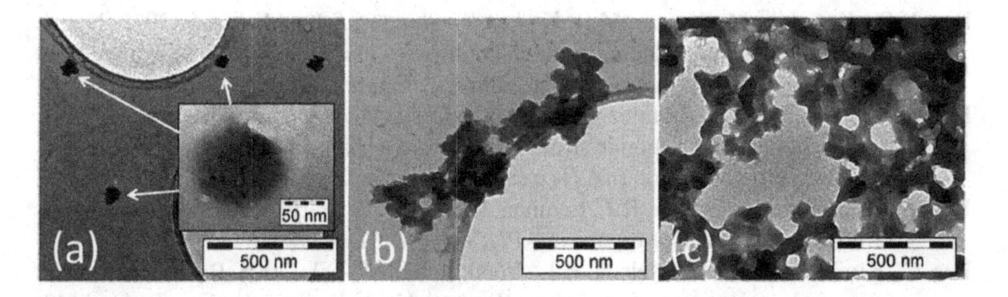

FIGURE 10.6 Morphologies of GT–Fe determined by TEM: (a) GT–Fe nanoparticles taken at (a) 1 min, (b) 60 min, and (c) 24 h of reaction time. [Reproduced with permission from Markova et al. (2014) © American Chemical Society].

includes exposure of three different chemical concentration iron oxide NPs suspension toward *Chlorella vulgaris* cells. Through investigation it was found that iron NPs impose oxidative stress, and inhibit cell division which further disrupts the photochemical activity of algal cells.

Similarly, in another study Markova et al. (2014) utilized ecologically important organisms such as cyanobacterium (*Synechococcus nidulans*), green alga (*Pseudokirchneriella subcapitata*), and invertebrate organisms (*Daphnia magna*) to study the effect of green synthesized iron nanoparticles on them. Experimental results conveyed that green synthesized iron nanoparticles obtained by using green tea extract induced a negative effect on toxicological assays prepared. Though studies indicate that green synthesized NPs are biocompatible to living organisms, there is still a lack of literature studying its toxicological effects on the environment (Figure 10.6).

REFERENCES

Antisari, L.V., Carbone, S., Gatti, A., Vianello, G., Nannipieri, P. 2013. Toxicity of metal oxide (CeO_2, Fe_3O_4, SnO_2) engineered nanoparticles on soil microbial biomass and their distribution in soil. *Soil Biol. Biochem.* 60, 87–94.

Auffan, M., Achouak, W., Rose, J., Roncato, M.-A., Chanéac, C., Waite, D.T., Masion, A., Woicik, J.C., Wiesner, M.R., Bottero, J.-Y. 2008. Relation between the redox state of iron-based nanoparticles and their cytotoxicity toward *Escherichia coli*. *Environ. Sci. Technol.* 42, 6730–6735.

Barhoumi, L., Dewez, D. 2013. Toxicity of superparamagnetic iron oxide nanoparticles on green alga *Chlorella vulgaris*. *BioMed. Res. Int.* 2013, 1–11.

Baumann, J., Koser, J., Arndt, D., Filser, J. 2014. The coating makes the difference: Acute effects of iron oxide nanoparticles on *Daphnia magna*. *Sci. Total Environ.* 484, 176–184.

Blinova, I., Kanarbik, L., Irha, N., Kahru, A. 2017. Ecotoxicity of nanosized magnetite to crustacean *Daphnia magna* and duckweed Lemna minor. *Hydrobiologia* 798, 141–149.

Chen, P.J., Wu, W.L., Wu, K.C. 2013. The zerovalent iron nanoparticle causes higher developmental toxicity than its oxidation products in early life stages of Medaka fish. *Water Res.* 47, 3899–3909.

Das, K.R., Kowshik, M., Kumar, M.P., Kerkar, S., Shyama, S.K., Mishra, S. 2018. Native hypersaline sulphate reducing bacteria contributes to iron nanoparticle formation in saltpan sediment: A concern for aquaculture. *J. Environ. Manage.* 206, 556–564.

Fajardo, C., Sacca, M.L., Martinez-Gomariz, M., Costa, G., Nande, M., Martin, M. 2013. Transcriptional and proteomic stress responses of a soil bacterium *Bacillus cereus* to nanosized zero-valent iron (nZVI) particles. *Chemosphere* 93, 1077–1083.

Keller, A.A., Garner, K., Miller, R.J., Lenihan, H.S. 2012. Toxicity of nano-zero valent iron to freshwater and marine organisms. *PLoS One* 7 (8), e43983.

Lee, C., Kim, J.Y., Lee, W.I., Nelson, K.L., Yoon, J., Sedlak, D.L. 2008. Bactericidal effect of zero-valent iron nanoparticles on *Escherichia coli. Environ. Sci. Technol.* 42, 4927–4933.

Li, H., Zhou, Q., Wu, Y., Wang, T., Jiang, G. 2009. Effects of waterborne nano-iron onmedaka (*Oryzias latipes*): Antioxidant enzymatic activity, lipid peroxidation and histopathology. *Ecotoxicol. Environ. Saf.* 72, 684–692.

Li, Z., Greden, K., Alvarez, P.J.J., Gregory, K.B., Lowry, G.V. 2010. Adsorbed polymer and NOM limits adhesion and toxicity of nano scale zerovalent iron to *E. coli. Environ. Sci. Technol.* 44, 3462–3467.

Markova, Z., Novak, P., Kaslik, J., Plachtova, P., Brazdova, M., Jancula, D., Siskova, K.M., Machala, L., Marsalek, B., Zboril, R., et al. 2014. Iron (II,III)—Polyphenol complex nanoparticles derived from green tea with remarkable ecotoxicological impact. *ACS Sustain. Chem. Eng.* 2, 1674–1680.

Nadagouda, M.N., Castle, A.B., Murdock, R.C., Hussain, S.M., Varma, R.S. 2010. In vitro biocompatibility of nanoscale zerovalent iron particles (nZVI) synthesized using tea polyphenols. *Green Chem.* 12, 114–122.

Phenrat, T., Long, T.C., Lowry, G.V., Veronesi, B. 2009. Partial oxidation ("aging") and surface modification decrease the toxicity of nanosized zerovalent iron. *Environ. Sci. Technol.* 43, 195–200.

Sacca, M.L., Fajardo, C., Costa, G., Lobo, C., Nande, M., Martin, M. 2014. Integrating classical and molecular approaches to evaluate the impact of nanosized zero-valent iron (nZVI) on soil organisms. *Chemosphere* 104, 184–189.

Shakibaie, M., Shahverdi, A.R., Faramarzi, M.A., Hassanzadeh, G.R., Rahimi, H.R., Sabzevari, O. 2013. Acute and subacute toxicity of novel biogenic selenium nanoparticles in mice. *Pharm. Biol.* 51, 58–63.

Usha Rani, P., Rajasekharreddy, P. 2011. Green synthesis of silver-protein (core-shell) nanoparticles using *Piper betle* L. Leaf extract and its ecotoxicological studies on daphnia magna. *Colloids Surf. A* 389, 188–194.

You, Y., Han, J., Chiu, P.C., Jin, Y. 2005. Removal and inactivation of waterborne viruses using zerovalent iron. *Environ. Sci. Technol.* 39, 9263–9269.

11 Critical Analysis and Future Scope of Green Synthesis Routes

11.1 CRITICAL ANALYSIS OF THE GREEN SYNTHESIS ROUTES FOR IRON-BASED NANOMATERIAL SYNTHESIS

Stability, size distribution, zeta potential, cost of preparation, and scaling up ability play important roles in the evaluation of the different green synthesis routes. Stability plays an important role where the NPs make metastable aqueous suspensions or aerosols when they enter into surface or water from the environment. It was found through several studies that plant extract and microorganism mediated green synthesis produced better stabilized NPs at low cost. Effective surface coating was formed due to the presence of various polyphenolic components in plant extracts, natural proteins, tannins, and so on in microorganisms, which provided better stabilization, and size distribution. Biocompatible reagents obtained from plant sources such as cellulose, clay, polysaccharides, and ascorbic acids are usually less expensive than sources like sugar, hemoglobin, synthetic tannins, and gallic acid. But such process-mediated nanoparticles provide better stabilization and less reactivity in the environment with narrow particle size distribution. The hydrothermal route is energy efficient, but requires low external power for heating and incurs high costs if chemical coating agents are required for better stabilization. Otherwise, if natural plant extracts are utilized as reducing agents, it can also be very economical. With better coating arises better stabilization which provides higher zeta potential indicating better dispersion in aqueous medium. Moreover, microwave-based green synthesis adopts the hydrothermal process and has proved to have better scalability properties. If microorganisms and plant extracts are used as the reducing agents then this route is a most economical way to produce NPs with better crystalline nature, stabilization, and size distribution. Moreover, green synthesized nanoparticles are found to be better dispersed and stabilized with higher zeta potential in aqueous medium at definite pH. The scale-up advantage persists for all the different green synthesis routes mentioned in this book, especially for the microwave-assisted technique. Usually, plant-based and microorganism-based green synthesis occurs at normal conditions and thus is regarded as less expensive, whereas the other synthesis routes require an external power source for heating and thermal decomposition for generating NPs.

DOI: 10.1201/9781003243632-11

11.2 SEPARATION STRATEGY OF MAGNETIC IRON NPs

Usually, in laboratory process magnetic separation is performed by low-gradient magnetic separation technique. In such process a low magnetic field gradient is generated (<100 T/m) by placing a permanent magnet adjacent to the sample, and separation of magnetic NPs takes place in batch mode. On the other hand, for industrial purposes high gradient magnetic separation is generally adopted. The setup consists of a column packed with magnetic NPs matrix composed of stainless steel wool. The NP matrix is left exposed to an external magnetic field. The suspension to be treated then flows continuously through the column. A large localized magnetic field gradient (~10^4 T/m) generated near the matrix surfaces serves to capture the magnetic materials from the flowing stream. On switching off, the external magnetic field releases the captured magnetic materials since the magnetic matrix gets demagnetized. Such process is successfully implemented in various chemical engineering plants such as food-processing, power plants, as well as wastewater treatment plants.

11.3 CHALLENGES AND FUTURE PROSPECTS OF GREEN SYNTHESIZED IRON-BASED NANOMATERIALS

More research and advancements are required to explore the uniqueness of green synthesized nanomaterials, and various other available resources for such synthesis should be trialed. Understanding the underlying mechanism of such synthesis plays an important role in discovering new modifications and advancements. Moreover, developments should lead toward an economically competitive process, can be scaled up, and should be eco-friendly in nature compared to conventional methods. Future studies should involve the use of more local resources to make things economically viable, and research should focus more on the stability issues of synthesized nanomaterials for better biocompatible applications. Research should put emphasis on synthesizing such green nanomaterials toward better efficiency in tackling contaminants with minimum eco-toxicological effects. Again, attention should focus more on discussing and controlling the risk management, fate, and toxicological effects of green synthesized iron nanoparticles concerning chemical synthesis.

a. It has been noted that some researchers have pointed out the issue of food security, regarding plant products being used as a source of reducing agents in green synthesis. Hence, more attention should be given to agro/biowaste as well as indigenous herbs to overcome this. Moreover, extensive studies are required regarding the fate and transport of nanoparticles in the environment.
b. The mechanism of reaction for the biosynthesis of nanoparticle varies according to different microorganisms such as bacteria, fungi, or yeast. Nanoparticles with varying morphology and characteristics arise due to the diverse reactions between metal ion precursors and different biological

agents. Intracellular synthesis involves bioreduction of metal ions on the cell wall, whereas extracellular synthesis occurs through electron transfer between enzymes and metal ions. Since such reactions are very complicated, intensive research needs to be carried out for a better understanding of such process mechanisms, which can cater to effective control over morphology and other properties of synthesized nanoparticles.

c. Iron oxide nanoparticles are reactive in nature, specifically zerovalent irons due to their higher oxidation state. Such nanomaterials during transport from one medium to another can cause serious toxic effects on soil and aqueous systems, which can negatively affect the ecosystem. Thus research should be focused on better stabilization of prepared nanomaterials through green routes, and better management of nanomaterial waste.
d. Research should be oriented toward utilizing agro-based waste for the green synthesis of nanomaterials. Moreover, metal-metal nanocomposite synthesis through green route should be more focused to enhance effectivity in wastewater treatment, and heavy metal removal application purposes.
e. Through this study it has been observed that different parts of the same plant species produce nanoparticles with different morphology and physic-chemical properties. Research should be encouraged to deal with a thorough study of the active biomolecules primarily responsible for capping and stabilizing the synthesized nanomaterials.
f. Exploration of new microorganisms is required for synthesis of iron nanoparticles. From past literature it has been observed that microorganism-related synthesis forms less stabilized and narrower morphological oriented nanomaterials, resulting in lower yield value. To obtain nanoparticles of uniform, crystalline, and monodisperse properties, new microorganism research should be focused on exploring different species.
g. To control the risk management due to the toxic nature of synthesized iron nanoparticles there should exist a standard method for the synthesis. Since different synthesis methods adopted by researchers lead to different property and characteristic oriented nanomaterials, proper screening of the nanowaste before discarding is often overlooked. This creates inconsistency in research output and difficulties in human health predictions along with environmental risk assessments.
h. Bimetallic nanoparticles are found to be a very interesting research area which tends to improve the properties of Fe NPs alone but is scarce in literature. Such bimetallic nanoparticles can improve the catalytic property, surface area for adsorption, and other various chemical properties. Research should be more focused on such bimetallic nanocomposites for various environmental applications.
i. The real challenge of utilizing such NPs as adsorbent in column operation is the huge pressure drops which build up due to exceptionally reduced bed void volume. Such challenges could be handled by incorporating NPs over support such as adsorbent beads, carbon composites, and activated carbon. For such operation chemisorption should be the ideal interaction between

the adsorbate and adsorbent. Hence preparing and choosing particular NPs with better surface functional groups would be the main priority.

11.4 CONCLUSION

The chapters of this book emphasize the various methods of green synthesis of iron-based nanomaterials (nanoparticles and nanocomposites) and their potential application toward environmental remediation. An effort is made to categorize various raw materials under green synthesis and analyze their respective application toward various environmental-related applications. The analysis made here brings forward minute details of nanomaterial size, shape, properties, and application of iron NPs. Various plant sources and biocompatible reagents along with energy efficient methods have been summarized in detail for preparing iron-based nanomaterials and nanocomposites. The numerous roles of iron nanoparticles as catalyst, adsorbent, and their toxicological effects are discussed in the book. Moreover, iron bimetallic nanocomposite synthesis through such a green route, which is evolving nowadays, has been pointed out. Plant-based synthesis has been compared with graphene-based composites due to their surface coatings with numerous plant metabolites, which act as excellent adsorbent. Moreover, critical analysis of the different green synthesis routes with respect to their stability, size distribution, and various other parameters has been evaluated. Finally, the challenges and prospects of green synthesis of iron NPs are discussed. In summary, various green nanotechnology processes have been described here, which would be helpful in future study as a resource to analyze and utilize for further advancements in building better iron-based nanomaterials to tackle real-life wastewater and contaminated solutions with least risk of a toxicological effect on the environment. This book might be useful to readers wishing to acquire in-depth knowledge on green synthesis of iron NPs and its amazing success for various environmental applications.

Index

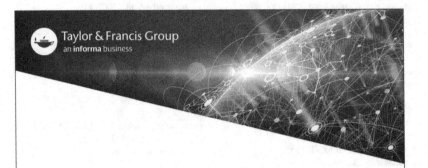

Printed in the United States
by Baker & Taylor Publisher Services

Printed in the United States
by Baker & Taylor Publisher Services